Vitamin C in Human Health and Disease

Wang Jae Lee

Vitamin C in Human Health and Disease

Effects, Mechanisms of Action, and New Guidance on Intake

Springer

Wang Jae Lee
Department of Anatomy
Seoul National University College of Medicine
Seoul, South Korea

ISBN 978-94-024-1711-1 ISBN 978-94-024-1713-5 (eBook)
https://doi.org/10.1007/978-94-024-1713-5

This Springer imprint is published by the registered company Springer Nature B.V.
The registered company address is: Van Godewijckstraat 30, 3311 GX Dordrecht, The Netherlands

Preface

More than 30 years have been passed since I first made vitamin C part of my life. I think it was my great good fortune to be introduced to vitamin C as a medical doctor, for I have been able to lead a healthier life by regularly taking large amounts of vitamin C. In addition, I have performed countless experiments on vitamin C using animals, especially gulo(−/−) mice, which, like humans, cannot synthesize this vitamin. These experiments, published in some 40 papers in recognized international journals over the last few decades, have served to identify the effects of vitamin C on each of the human body systems, including the cardiovascular, nervous, gastrointestinal, immune, endocrine, and respiratory systems, and have sometimes even elucidated the detailed mechanisms involved.

In most scientific books, chapters are co-authored, but I have chosen to write this book as the sole author because I dare to declare myself a "vitamin C specialist". Moreover, throughout my career I have lectured on the structure of the human body as an anatomy professor, and I consequently believe that I have a deep understanding of the whole human body.

I do not regard this book as purely scientific. Medicine is an empirical science. In the treatment of the sick, endless scientific endeavors aim to clarify the treatment mechanisms, but in many cases the treatment is successful without knowing the exact mechanism of action. Indeed, it is virtually impossible to reveal the mechanisms of all treatments for all diseases because clinical experimentation using humans is extremely limited. The purpose of the book is, therefore, not merely to assemble the relevant scientific evidence but to enhance the health of people across the world by explaining why vitamin C should be taken regularly and by describing the precise amount of vitamin C to take each day and how to take it.

As I have just mentioned, this book is intended to promote human health through the proper use of vitamin C. In this context, each body system/organ-oriented chapter begins with a very brief description of structure because understanding of the human body must precede discussion of measures to benefit human health. Thereafter, the effects of vitamin C on each system or organ are described based mainly on the results that I have obtained during my studies. Of course, I have done my utmost to ensure that the scientific basis for taking vitamin C is sound by fully reviewing and taking into account the research results of other international scientists.

I want to let every reader know that the most important point of the book is the presentation of a new proposal on *the optimal dose of vitamin C,* which is currently 60 mg or more. Although the recommended dose has increased over the years, for too long insufficient detailed consideration has been devoted to the actual importance of vitamin C in maintaining good health. Furthermore, I wish to emphasize that the purpose of regular vitamin C use is not merely to prevent scurvy, but to actively prevent fatal diseases such as cancer, vascular diseases, and so on, which have become the fate of so many people in the modern world.

Above all, I want to glorify God, who has deepened enlightenment and inspiration during my writing of this book and share joy with my family, wife and daughter. And then, I would like to express my sincere gratitude to the many people who have supported me through their continuous prayers for successful publication. Especially, CEOs, Changhyon Cho and Byung-Jo Yim, who supported me with the grant for the research of vitamin C as well as helped me publish the book of vitamin C for the first time in my life. Furthermore, I cannot help but notice that there were so many people helping me with this book. Professor Kwang-Ho Lee, who informed me of the importance of vitamin C for the first time more than 30 years ago; Dr. Daeho Cho, who performed vitamin C research in my lab for the first time; Drs. Dong-Hoon Jin and Jae Seung Kang, who have greatly improved and enriched vitamin C research both quantitatively and qualitatively, providing many scientific data referred to in the book; all my beloved PhD students, who have contributed to the study of vitamin C without complaint; Mr. Yong Ho Yeo, CEO of Paramark, and Mr. Sae Young Chung, who helped to edit the figures for the book; Ms. Sunjoo Kim, who made many of the illustrations by herself; and Ms. Young-jin Oh, my secretary, who has handled all the prepublication chores, including compilation of the manuscripts and procurement of permission to use illustrative material from other publications. I also express my thanks to Ms. Yim, the former publishing editor, and Dr. Sue Lee and Ms. Emmy Lee, the publishing editors of Springer, Biomedical Sciences for their active support.

Finally, I am sincerely grateful to the many people who have listened to my recommendations through my lectures, the mass media, or social network sites directly or indirectly, waiting for this book to be published while taking a high dose of vitamin C over the past 30 years.

Seoul, South Korea Wang Jae Lee

Contents

Chapter 1
Introduction

History of Vitamin C Discovery

There have been many episodes of vitamin C deficiency, known as scurvy, throughout human history, for example, a scurvy-like disease written down on papyruses (1550 BC) and a Greek physician's (Hippocrates) description of scurvy-like symptoms (400 BC); however, only the core episodes that provide living evidence on the importance of vitamin C in maintaining life are summarized and described chronologically in this chapter (Table 1.1).

What Is Vitamin C?

Vitamin C is a representative of the water-soluble vitamins. Another representative of that group is vitamin B, which now is known to be a complex composed of several subfractions, such as B6, B12, riboflavin, niacin, biotin, folic acid, and so on. Considering that both vitamins are commonly water-soluble, what are the basic differences between vitamins B and C? There is no doubt that they must have their own functions. Traditionally, vitamin B deficiency has been called beriberi, while vitamin C deficiency has been called scurvy. The interesting difference between them is that people rarely die of beriberi as long as they are normally fed in their routine life. Actually, no reports on deaths from vitamin B deficiency have been documented in human history because they usually die of starvation in advance just before they die of beriberi. However, in the case of vitamin C, it is very different. People can die of vitamin C deficiency, and actually, many people have died from it even though they are regularly fed. The medieval history of Europe and especially the British marine history, during which people could not help but use marine transportation for long-distance travel, tell us that many young marine soldiers died without any cause during the voyage to the land under British occupation even though

W. J. Lee, *Vitamin C in Human Health and Disease*,
https://doi.org/10.1007/978-94-024-1713-5_1

Table 1.1 Brief chronicle on vitamin C discovery

1550 BC. A scurvy-like disease, thought to be treated by eating onions, is written down on papyruses
400 BC. Hippocrates, a Greek physician, known to be the "father of modern medicine" describes scurvy-like symptoms: fetid breath, lax gums, and bleeding from the nose
1499 AD. When Vasco da Gama came back to Lisbon, Portugal, from the Arabian Sea, he lost 116 out of 170 of his crew despite knowing that citrus fruits could treat scurvy
1535 AD. While struggling against scurvy caused by the extremely cold winter on the St. Lawrence River, Canada, French explorers Jacques Cartier and Daniel Knezevic were advised by native Indians to drink plant tea from the needles of the arbor vitae tree to prevent death from scurvy during the winter
1753 AD. James Lind, British naval surgeon, made the report "A Treatise on the Scurvy" after the first controlled in vivo experiment (1747), using 12 men, who were suffering from scurvy. They were divided into six groups, each of which were given different regimens with a basic diet. Among them, the citrus fruit feeding group showed the most remarkable recovery from scurvy
1795 AD. Gilbert Blane, a British naval physician, advised the admiralty to eat citrus fruits on board ships as a routine diet. Thereafter, scurvy has been forever eliminated from the British navy
1802 AD. The British government established a law which mandates every ship to load citrus fruits on board before any voyage for any reason
1880 AD. Since the establishment of the law mandating every sailor eat citrus fruits on board ships, British sailors had to eat citrus fruits on board ship to prevent scurvy. However, because they preferred lime among the citrus fruits such as lemons, oranges, and limes, British sailors were called "limeys"
1907 AD. Axel Holst and Theodor Frølich, two Norwegian physicians studying shipboard beriberi, had a serendipitous finding from guinea pigs in which the scurvy-like condition was caused when they were fed only food consisting of grains and flour and that condition was cured when the test animals were fed fresh vegetables. This became the starting point of intensive scientific research on scurvy
1912 AD. The vitamine is named by combining "vital" and "amines" because it was thought that all these materials were vital as well as chemical amines. Although the "e" was deleted after it was proven that some of these compounds are not amines, the word vitamin is used popularly as a generic name for them
1928 AD. Hungarian biochemist Albert Szent-Györgyi isolated an organic reducing substance, hexuronic acid, from animal adrenal glands and suspected it to be an antiscorbutic factor. However, he could not prove it to be an antiscorbutic factor
1932 AD. Charles Glen King isolated vitamin C in his laboratory at the University of Pittsburgh and concluded that it was the same as "hexuronic acid." However, he could not determine its chemical structure
1933 AD. British chemist Walter Haworth deduced the molecular structure of hexuronic acid and renamed it ascorbic acid
1937 AD. Albert Szent-Györgyi was nominated as a laureate of the Nobel Prize in Medicine for his discovery of vitamin C, and Walter Haworth shared that year's Nobel Prize in Chemistry for his work with carbohydrates and for his part on the vitamin C synthetic work
In 1970s AD. Linus Pauling, who is the only person ever to receive two unshared Nobel Prizes (Chemistry in 1954 and Peace in 1962), made vitamin C popular for health promotion and disease control, by writing books on the effects of vitamin C for health promotion including *Vitamin C, the Common Cold, and the Flu* and *Cancer and Vitamin C*

(continued)

Table 1.1 (continued)

End of 1970 AD. Linus Pauling established the Linus Pauling Institute of Science and Medicine, where many scientists continue to perform research on the roles of vitamin C most vigorously in the world
1994 AD. Linus Pauling died of pancreatic cancer. After his death, the Linus Pauling Institute of Science and Medicine moved to the campus of Oregon State University, his undergraduate alma mater

they were fed on a normal basis. At that time, most Europeans thought their deaths were a kind of sacrifice to the god of the sea. However, James Lind, a Scottish surgeon in the Royal Navy, did not agree with that idea and had kept thinking about the cause of their deaths. One day, when his ship had been sheltering from a hurricane on the coast of a northern African province during another voyage, he happened to find the cause of their deaths. He witnessed that many marine soldiers suffering from some kind of fatal illness had survived just after they ate some oranges or limes, which had been purchased from a merchant ship which was also sheltering from the hurricane. And since then, for the first time, that fatal disease was called scurvy, although vitamin C deficiency was proven to be the cause of that fatal disease much later on. He even did some human clinical trials using the marine soldiers on the voyage to prove his observations that he had made while sheltering at the African province. While at sea in May 1747, Lind fed some sailors citrus fruits (two oranges and one lemon) each day as well as their routine rations, while the others (actually five groups according to the different sorts of additional diets other than citrus fruits) continued on cider, vinegar, sulfuric acid, or seawater, along with their normal rations. The results conclusively showed that among them, the citrus fruit-fed group had the most remarkable recovery from scurvy; therefore, citrus fruits could prevent the disease. In the history of science, this should be considered the first recorded controlled clinical trial as well as a splendid achievement of science because his great scientific work reduced the number of victims that died from that fatal disease scurvy and even saved the lives of many people who should have died of scurvy otherwise. Lind published his work in 1753 in his *A Treatise on the Scurvy* (Lind J. *A Treatise of the Scurvy,* 1753). About 50 years later, at the beginning of the nineteenth century, the British government established a new marine regulation which permits the departure of any ship from a port only if that ship is loaded with limes or oranges based on the results of Lind's human trial. Thereafter, since the establishment of the law mandating every sailor eat citrus fruits while on board a ship, British sailors had to eat citrus fruits while on a ship to prevent scurvy. However, because they preferred limes among citrus fruits such as lemons, oranges, and limes, the British marine officers were nicknamed as "the limeys."

Why should there be such basic differences between scurvy and beriberi? The answer is because vitamin B is a real vitamin, while vitamin C is not a vitamin. By definition, a vitamin is a very small amount of a substance contained in food which humans or animals consume every day to maintain the health and functions of their body and to prevent death from starvation, in other words, to essentially keep on living. Hunger usually causes an increased appetite which makes the body eat whatever is available. Most of the food consumed to sustain life is composed of energy sources (carbohydrates and fats) because energy is essential for maintaining life because prolonged starvation will eventually lead to death. Therefore, many people, even children, say they are "starving to death" whenever they feel hunger. This relatively long explanation is the reason why any people cannot die of vitamin deficiency as long as they regularly consume some food which usually contains minerals in very small amounts including vitamins (A, D, E, F, and K), electrolytes (Na^+, Cl^-, K^+, Ca^{++}, and so on), and some metal ions (Fe, Zn, Mg, and Cu). Therefore, it is quite natural that minerals are commonly called "micronutrients." However, it is completely different in the case of vitamin C which is normally produced in the bodies of mammals except for primates which include humans. Specifically, vitamin C is produced in the liver. The raw material for vitamin C production is D-glucose which is converted into L-ascorbate through some enzymes, among which the last one is an oxidase for conversion of L-gulono-γ- lactone into L-ascorbate. The gene encoding the oxidase is mutated in primates, resulting in a failure of vitamin C production. The reason why these scientific facts about vitamin C are very important is that several conditions should be considered to determine the optimal dose of vitamin C to promote human health or to prevent disease which are different from those of other real vitamins. Nowadays, although many people take vitamin C worldwide and its market size is actually very large, there are still controversies surrounding Vitamin C. Although many specialists on vitamin C promote it widely, even laypeople know that vitamin C is very important in maintaining life and a healthy body. However, the biggest issue on vitamin C consumption has been always the daily recommended allowance.

Brief Explanation of the Biochemical Properties of Vitamin C

Many people have known that vitamin C is just one of many antioxidants, and some people even think that vitamin C is the strongest antioxidant. Those facts are not true. Actually, vitamin C has unique functions beyond just its function as an antioxidant. Here, three points should be stressed with regard to the real biochemical characteristics and biological importance of vitamin C. First, vitamin C is water-soluble. Water-soluble substances usually have a very easy accessibility to any cells or organs as well as have rapid reaction potentials. Therefore, vitamin C is superior in preventing oxidative damage to the body compared to other antioxidant vitamins such as vitamin A, vitamin E, or β–carotene, which are lipid soluble. Second, vitamin C has a very unique pattern of antioxidant action. Chemically, an antioxidant is a kind of electron donor so that it is usually converted into a radical which is very toxic to cells or tissues. Therefore, actually, in the body, radicals are usually removed within an ultrashort time period after their generation. If not, tissues and cells will be severely damaged by the radicals, which in some cases result in death. Vitamin C, L-ascorbate (reducing form of vitamin C), is converted into the L-ascorbyl radical after the first donation of one electron and then finally converted into dehydroascorbate (DHA, oxidizing form of vitamin C) after the second donation of one electron, so that vitamin C has a two-step action pattern for the antioxidant unlike other antioxidants. Furthermore, the potential of one-electron transfer is very low, implying that vitamin C can easily act as an antioxidant, but its antioxidant strength is weak. And in some cases, vitamin C has a prooxidant property. Especially, vitamin C must act as prooxidant in cancer cells. That is to say, even though vitamin C is known to be one of the many antioxidants, it has a unique redox potential. This is also the reason why vitamin C is known to be chemically very unstable. In the living body, there should be a key substance which can easily and harmlessly eradicate radicals within an ultrashort time period which are inevitably produced while preventing oxidative damage to tissues or cells by the antioxidant actions of so many antioxidants present in their own form. Very interestingly, vitamin C also generates radicals after its antioxidant action like other antioxidants, but the ascorbyl radical is the least toxic. The chemical instability of vitamin C is also the third biochemical characteristic of vitamin C. Actually, vitamin C has an important role in regenerating toxic radicals derived from many antioxidants after their action into the safe original form. For example, vitamin E is a well-known antioxidant vitamin, which is called α-tocopherol, and is known to have a role in preventing lipid peroxidation, which is a prerequisite condition for atheroma formation. α-Tocopherol is converted to the α-tocopheroxyl radical, which is very toxic to the body, after its antioxidant action, and then, the radical is regenerated into the original form, α-tocopherol by vitamin C within a very short time. Conclusively, there are clear reasons why vitamin C is water-soluble and has the least toxic radical after its action. There are many antioxidants that have their own roles in preventing oxidative damage to the body, especially in the deepest part of the body, and as such, many kinds of radicals are generated. Those radicals should be regenerated to the safe original form very quickly and without any damage to the body by the water-soluble and least toxic radical generating antioxidant, otherwise known as vitamin C. Therefore, these characteristics of vitamin C should be considered before people take other antioxidant vitamins. Namely, they should take an excessive amount of vitamin C so that their body is fully saturated with vitamin C before they take other antioxidant vitamins, such as vitamin A, vitamin E, and β-carotene. Full knowledge of these characteristics of vitamin C should help to understand the Copenhagen report (called "Copenhagen shock" because the report was shocking to general people) published in JAMA (Dr. Chritian Gluud, 2007.) (Refer to the next session, *Copenhagen Report*).

Copenhagen Report

This report has been also called "Copenhagen shock" because the facts included in this report were shocking to many people in the world. This report has been published in JAMA (February 28, 2007) with the title of the article "Mortality in Randomized Trials of Antioxidant Supplements for Primary and Secondary

Prevention: Systematic Review and Meta-analysis," performed by European scientists (the principal investigator is Dr. Goran Bjelakovic) affiliated with several institutes such as the Cochrane Hepato-Biliary Group, Cochrane University Hospital, Department of Internal Medicine at the University of Nis, and Divisione di Medicina in Palermo, Italy. They analyzed the data statistically through systemic review of many reports and then meta-analysis for 3 years and 3 months in 68 independent scientific articles from 232,606 participants (male, 55.5%; female 44.5%), who have taken antioxidant supplements daily or once in a while for 2 years and 7 months (average); their physical changes have been checked. The results were contrary to the concepts conventionally accepted as a health dogma, in which antioxidants can give the beneficial effects on human health. According to the summarized results, the combination treatment with beta-carotene, vitamin A, and vitamin E increases mortality by 5%, and independent treatment with vitamin A increases mortality by 16%, with vitamin E by 4%, and with beta-carotene by 7%, respectively. And the potential roles of vitamin C and selenium need further study (Bjelakovic et al. 2007). One year later (April 2008), a very similar report, entitled "Antioxidant Supplements for Prevention of Mortality in Healthy Participants and Patients with Various Disease (Review)," has been published in the latest issue of *The Cochrane Library* by Dr. Christian Gluud, director of medical science, associate professor and department head of the Copenhagen Trial Unit at the Centre for Clinical Intervention Research and Copenhagen University Hospital in Denmark, and his colleagues. After reviewing nearly 70 randomized studies on the effects of vitamin or antioxidant supplements, they concluded there is no evidence that they prolong life and even found evidence that some of them, such as beta-carotene, vitamin A, and vitamin E, may actually shorten it (Bjelakovic et al. 2008).

These reports have made many average people get in deep confusion and panic because they have known and believed that antioxidant supplements can scavenge oxygen free radicals so that body damages by them could be reduced, resulting in successful maintenance of their health. Even there have been so many debates on these shocking reports. Although these reports do not seem to be absolutely correct, definitely they look like proving a part of truth about the role of antioxidant supplements in keeping sound human health.

In this chapter, a couple of critiques on these reports will be given. First of all, it should be noted that there are no evidences in these reports that vitamin C might have the harmful effects on human health when they take vitamin C for a long time. Rather, among the reports, the fact that vitamin C could reduce mortality rate by 12% has not been known well to the public. It is more important to understand the reason why there have been few reports that vitamin C could prolong life. Because in every trial using vitamin C, people take too small amount of vitamin C (maximum 500 mg daily). Even worse, they do not abide by the correct method of administration, which is to take vitamin C every 6 h, considering the pharmacokinetics of vitamin C in the blood. Secondly, in the case of beta-carotene, it is quite natural that beta-carotene could increase mortality by 7% because the participants took six times of beta-carotene (30 mg) as much as recommended optimal daily dose (5 mg). Lastly, most scientists, who performed the experiments for the data included in the

Copenhagen report, do not look like they have an exact knowledge of biochemical properties of antioxidants. Generally, every antioxidant has become toxic radicals except vitamin C after functioning as antioxidant. For example, vitamin E, also called chemically alpha-tocopherol, which is a stronger antioxidant than vitamin C, becomes alpha-tocopheroxyl radical, which is very toxic to the body and can give chronic damages to cells and tissues within a very short period, on functioning as potent antioxidant, thereby known to prevent lipid peroxidation, resulting in prevention of vascular atherosclerosis. Therefore, regeneration of toxic tocopheroxyl radical to nontoxic tocopherol should be needed urgently in the body. In this regeneration process, vitamin C has been known to play a very important role as regenerator. Interestingly, only ascorbyl radical, which is a by-product formed after first step of antioxidant action (one-electron donation), does not have toxicity enough to give the damage to cells and tissues. In addition, because vitamin C is water-soluble, it is quite natural to have very quick and broad antioxidant function and have no potent adverse effects. It is surely understandable that the phenomenon of Copenhagen shock could never happen in the animals, which can synthesize substantial amount of vitamin C daily in their body. This is a reason why people should benchmark the animals when they want to keep their health by administration of antioxidant supplements. Conclusively, people should take adequate amount of vitamin C three times daily (i.e., every 6 h) and then even better to take other antioxidant supplements because fully saturated vitamin C in the body can prevent potential toxic effects caused by antioxidant radicals .

Other Biological Activities of Vitamin C Besides Its Roles as an Antioxidant

Vitamin C has very important functions beyond the biological roles as a key antioxidant. Those roles are as important cofactors of several enzymes involved in key processes of life. Classically, a representative symptom of vitamin C deficiency is delayed wound healing, which requires synthesis of an adequate amount of procollagen, which needs the maximum activity of two enzymes, (1) prolyl and (2) lysyl hydroxylases (Levine 1986).

Vitamin C has a very important role in the biosynthesis of carnitine, which has a critical role in energy metabolism by transporting long fatty acids into the mitochondria for β-oxidation, in which the maximum activities of (3 and 4) two dioxygenases are essential (Burri and Jacob 1997). Actually considering that the early symptoms of vitamin C deficiency are fatigue and lethargy (Levine et al. 1996b), it is not surprising that some people usually confess the termination of these early vitamin C deficiency symptoms after taking a substantial amount of vitamin C regularly.

Vitamin C has also a critical function in the biosynthesis of norepinephrine, which is a representative stress hormone. Vitamin C has a role as a co-substrate for

(5) dopamine β-monooxygenase for dopamine hydroxylation by converting dopamine to norepinephrine (Levine 1986; Burri and Jacob 1997; May et al. 2013). Not surprisingly, the concentration of vitamin C in the adrenal glands, where the most important stress hormones epinephrine (adrenaline) and norepinephrine (noradrenaline) are normally produced in the medullary zone, is 100–200 times higher as that in the peripheral blood. In addition, it is quite natural that vitamin C is usually depleted in the peripheral blood after prolonged exposure to mental or physical stress, which requires more stress hormone production.

Vitamin C is also known to be involved in cholesterol metabolism. Cholesterol is indispensable for the production of steroid hormones including sex hormones, which have critical functions in energy metabolism, the maintenance of blood pressure, the inflammatory process, and so on. In addition, bile acids, which are synthesized in the liver and help lipid digestion by emulsifying lipids in the small intestine, are formed from cholesterol. During these two metabolic processes, vitamin C has a critical role as a co-substrate of (6) the enzyme 7α-monooxygenase (Burri and Jacob 1997).

Vitamin C is also well known to be involved in downregulating high blood pressure by increased endothelial nitric oxide (Duffy et al. 1999), mediated by the enhanced activity of (7) endothelial nitric oxide synthase by keeping its cofactor tetrahydrobiopterin in the active, reduced form (Huang et al. 2000; Gewaltig and Kojda 2002).

Vitamin C is also known to increase the activity of a family of (8) oxygen-dependent prolyl hydroxylase enzymes only in vitro, which plays an important role in the ability of cells to recognize and respond to hypoxia (Bruick and McKnight 2001; Jaakkola et al. 2001; Vissers and Wilkie 2007; Page et al. 2008; Kuiper et al. 2014), but my research team reported that vitamin C also had the same function in vivo (Li et al. 2010).

Sodium-Dependent Vitamin C Transporters (SVCTs)

Understanding how vitamin C can be transported into and out of the cells should be discussed first before a review of the functions and biological significance of vitamin C. Up to the twentieth century, it has been known that vitamin C can be transported only through several glucose transporter proteins (GLUTs), which can mediate transportation of only the oxidized form of vitamin C (dehydroascorbate, DHA), which is immediately converted into the reducing form (L-ascorbate) chemically by the action of glutathione (GSH) or enzymatically by several catalases within the cell. In the peripheral blood, almost 98% of vitamin C is present in the reducing form (L-ascorbate), while a very small proportion (less than 2%) is present in the oxidizing form (DHA). How is the reducing form of vitamin C transported into the cells? In 1999, Dr. Hediger's group has first cloned the SVCT genes and reported the full sequence of the genes (Tsukaguchi et al. 2007), which made it possible to know that the trans-plasma membrane gradient of ascorbate is generated by specific ascorbate transporter proteins, two isoforms of which are known, SVCT1 and SVCT2. While SVCT2 is distributed widely in several organs such as the placenta, liver, retina, spleen, endocrine organs (adrenal glands, prostate, testis, and ovary), and brain (neurons), where the tissue concentration of vitamin C is exceptionally high, reaching up to 10 mM, SVCT1 is distributed in only some limited organs, such as the epithelial surfaces of the intestine and the proximal tubules of the kidney, where vitamin C is actively absorbed or reabsorbed, suggesting that SVCT1 is involved in the bulk transport of vitamin C, whereas SVCT2 is involved in organ- or tissue-specific uptake of vitamin C (Hediger 2002). The distribution pattern of SVCT2 in several organs or tissues reflects the specific role of vitamin C in each organ. For example, the concentration of vitamin C in the adrenal gland, with one of the highest expressions of SVCT2, where several steroid hormones such as glucocorticoid, mineralocorticoid, and even sex hormones (estrogen progesterone and testosterone) are produced via biochemical pathways, in which vitamin C has a key role as a cofactor for some enzymes, is 10 mM. Another good example is the brain, which is composed of neurons and supporting cells such as astrocytes and glia. The brain consumes 20% of the total oxygen inhaled despite its low proportion (2%) of total body weight. The amount of oxygen consumption by neurons is five times higher as those of the supporting cells. The intracellular concentration of vitamin C in neurons is 10 mM, whereas those in the supporting cells are 2 mM, indicating that the vitamin C concentration is proportional to the amount of oxygen consumption in each organ. Therefore, it is not surprising that the expression of SVCT2 in neurons, which use a greater amount of oxygen than other organs or cells, is at the highest level in the body.

Pharmacokinetics of Vitamin C

There are three active forms of vitamin C in the biological system: ascorbate (reduced form), ascorbyl radical (intermediate oxidized form), and dehydroascorbate (DHA, oxidized form) (Fig. 1.1).

Ascorbate is mainly absorbed in the small intestine actively through SVCT1 expressed on the apical side of enterocytes. However, the transportation mechanism from enterocytes to the blood vessels (the so-called basolateral efflux) is not known yet. The physiological characteristics of the intestinal absorption of DHA in humans, which is known to be mediated via GLUTs, especially GLUT1, GLUT3, and, to some extent, GLUT4, are not well known yet (Levine et al. 1996a, b; Rumsey and Levine 1998). Intestinal absorption of ascorbate and DHA is known to be inhibited by glucose in a competitive inhibition manner caused by the structural similarity between both substances (Malo and Wilson 2000).

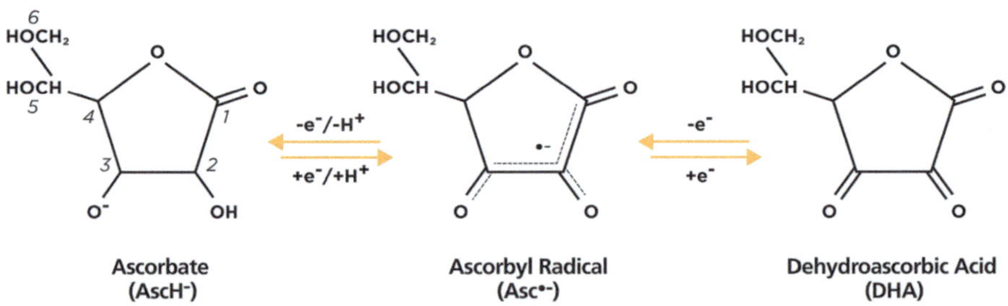

Fig. 1.1 Three forms of active vitamin C. Oxidation of ascorbate (AscH-) by two serial one-electron donation steps to give the ascorbyl radical (Asc・ -) and dehydroascorbic acid (DHA), respectively. Due to low one-electron transfer potential, the two-way reactions among three forms occur very easily, which is the reason why vitamin C has been known to be chemically unstable (Higdon and Frei 2002)

Before comments about the absorption, distribution, and excretion of vitamin C, it is necessary to understand actually how the bioavailability of vitamin C works in living organisms. In pharmacokinetic terms, bioavailability means the difference between the blood levels of vitamin C administered orally and intravenously. In other words, it reflects the absorption rate of vitamin C administered orally. As the only study showing the true bioavailability of vitamin C in humans, the bioavailability was very diverse according to the amounts of vitamin C given orally: when a liquid solution of ascorbate was given to fasting men at a steady state, the bioavailability was more than 80% for a dose less than 100 mg, 78% for a 200-mg dose, 75% for a 500-mg dose, and 62% for a 1250-mg dose (Graumlich et al. 1997). Two pharmacokinetic studies using healthy young men and women showed that doses of vitamin C ranging from 30 to 100 mg/day increased the blood concentration of vitamin C rapidly, whereas doses more than 400 mg/day of vitamin C showed a decreased rate in its increased blood concentration (Levine et al. 1996a, 2001). These reports have usually provided scientific rationale for the determination of the optimal daily allowance of vitamin C. When considering several characteristics of vitamin C, the blood level of vitamin C can not only be a guideline for determining the optimal daily allowance. About this issue, further discussion will be reviewed in the next step, "Scientific Background on the Optimal Dose of Vitamin C."

Details on the urinary excretion of vitamin C should be scrutinized because the excretion pattern is one of the several important factors to consider when determining the optimal daily allowance of vitamin C. Actually, urinary excretion of vitamin C will be discussed thoroughly in the next step, "Scientific Background on the Optimal Dose of Vitamin C."

The usual vitamin C concentration in human peripheral blood ranges from 11 to 90 μM. People whose plasma concentration is less than 11 μM are called scurvy patients. Those people whose plasma concentration range is from 11 to 28 μM are in a sub-scurvy status (Jacob 1999).

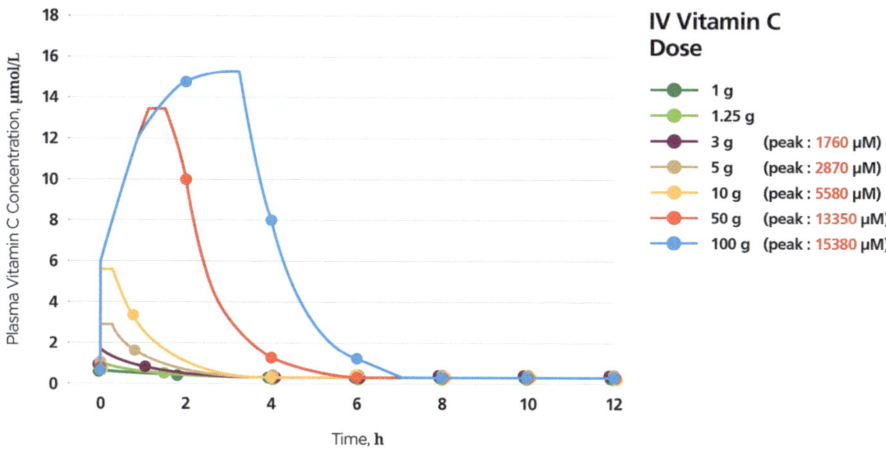

Fig. 1.2 Predicted plasma vitamin C concentrations in healthy persons after intravenous administration of vitamin C. In the case of intravenous administration, the blood concentration of vitamin C is increased in proportion to the dose administered at low concentration, but the degree of increase is slowed at the high concentration over 50 g so that the peak of 50-g administration is 13.35 mM, but the peak of 100 g administration is 15.38 mM, not high as twice as that of 50-g administration (Padayatty et al. 2004)

It is very important to have knowledge about the pharmacokinetics of vitamin C for its intake because it is critical to determine the optimal dose as well as its administration route. Theoretically, IV administration could be expected for its pharmacokinetics because the facts related to absorption in the GI tract do not have to be considered. Actually, IV administration shows an immediate peak of plasma vitamin C within 30 min. after administration. The actual plasma concentrations are dependent on the amounts of vitamin C administered. IV injections of less than 10 g reach peak within 30 min., while injections of 50 and 100 g show a peak at 2 and 4 h, respectively, after injection. However, their plasma levels go down to the baseline level abruptly within a few hours after injection (Fig. 1.2).

However, oral intake shows a different pattern of plasma levels. Administration of less than 500 mg keeps a constant plasma level of 80–90 μM for several hours, while the administration of more than 1 g of vitamin C shows that the peak levels go up as the amount of intake goes up and ultimately end up at less than 250 μM regardless of the amount administered (from 1 to 18 g). Single intakes show peak levels at 2–3 h after intake and go back to the original levels at 6–10 h after intake. Multiple intakes, for example, every 4 or 6 h, constantly keep relatively high levels (about 220 μM) (Fig. 1.3).

The type of vitamin C products changes the pharmacokinetics. The tablet type of vitamin C reaches its peak plasma level at 3 h after oral intake, while the aqueous product shows its peak plasma level at 2 h after oral intake (Fig. 1.4).

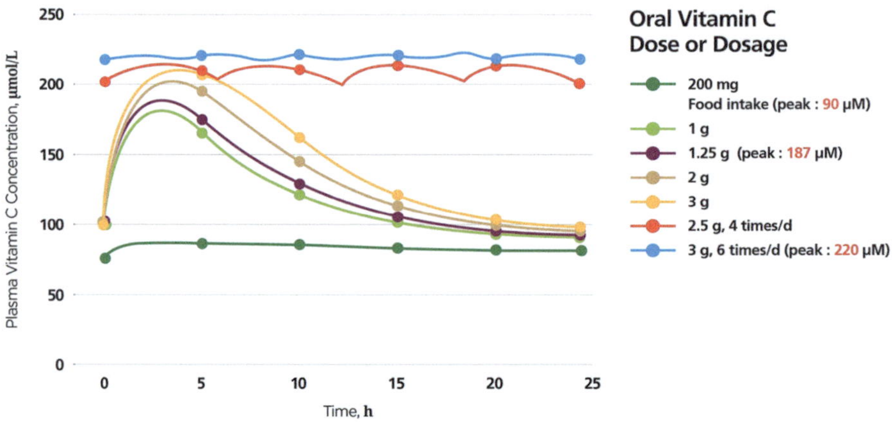

Fig. 1.3 Predicted plasma vitamin C concentrations in healthy persons after oral administration of vitamin C. In the case of oral administration, the plasma concentration of vitamin C increases to some extent as the dose increases, but the blood concentration of it never exceeds 250 M even if the doses are increased further. And approximately 3 h after oral administration, the peak blood levels are reached, and the concentration returns to its original level after 6–10 h. So, when vitamin C is administered every 4 or 6 h, the peak levels are maintained (Padayatty et al. 2004)

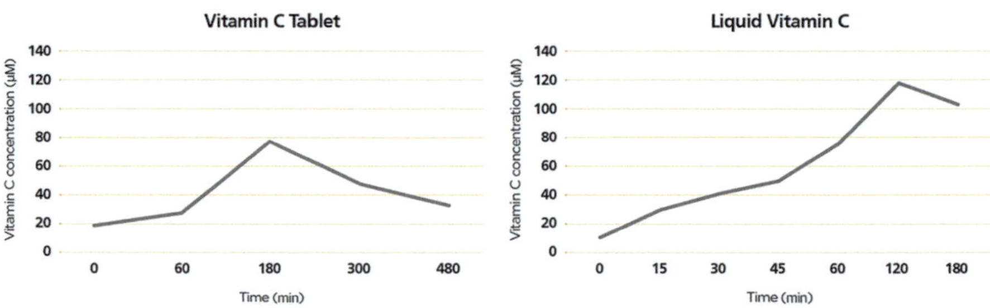

Fig. 1.4 Comparison of plasma vitamin C concentrations in healthy persons after oral administration of two types of vitamin C products, tablet and liquid products (unpublished our data). Liquid product reaches higher peak more quickly than tablet one (unpublished our data)

Scientific Background on the Recommended Daily Allowance (RDA) of Vitamin C

The actual optimal dose of vitamin C has been kept at range of 60~100 mg per day worldwide because before 2000, the RDA for vitamin C was based on scurvy prevention with a margin of safety (Food and Nutrition Board, 1980, 1989). However, there has been a recent trend in the RDA gradually increasing in many countries (German Society of Nutrition, Austrian Society of Nutrition, Swiss Society of Nutrition Research, Swiss Union of Nutrition 2000; Health and Nutrition Information Society, Tokyo 2000; Birlouez-Aragon et al. 2000). In the USA, the Food and

Nutrition Board of the US National Academy of Sciences as part of new Dietary Reference Intake guidelines reported in 2000 that the optimal doses for men and women were 120 and 100 mg, respectively, and even declared that a daily intake of up to 2000 mg of vitamin C would be tolerable (Yates et al. 1998; Food and Nutrition Board, Panel on Dietary Antioxidants and Related Compounds 2000). Historically, in 1970, Dr. Linus Pauling, one of the most conspicuous frontier scientists on vitamin C, had deduced that the optimal daily dose of vitamin C would be at least 2300 mg/day for an adult with an energy requirement of 2500 kcal/day, by considering facts such as the daily production amount of vitamin C by several animals, evolutionary speculations, and the amount of vitamin C consumed in a raw plant food diet during actual human life (Pauling 1970). To our regret, even though his evolutionary opinion on the daily optimal dose of vitamin C has not been widely accepted in the medical community due to a lack scientific evidence, it has had a great impact on the general population with a persistent increase of vitamin C consumption by the general population. Actually, vitamin C products, for which the vitamin C contents in a tablet is 1000 mg, have been released like a flood in many markets in the USA, even though the official optimal dose of vitamin C is still much less than 500 mg. What are the rationales for sustaining the optimal dose (60 mg per day) for a long time? The first rationale on which the dose is determined is if there are symptoms of scurvy, which are weakness, anemia, spongy gums, a tendency for mucocutaneous hemorrhages, and a brawny induration of the muscles in the calves and legs, as well as death which is a terminal symptom of an extreme deficiency case. In reality, if people keep taking 60 mg of vitamin C every day, no symptoms of scurvy occur. The second one is the urinary excretion pattern after oral intake of vitamin C. Within 3 h after oral intake of 60 mg vitamin C, a very small amount of vitamin C is excreted in the urine, which is usually interpreted in terms of pharmacokinetics that more than 60 mg of vitamin C is not needed in the body. Our unpublished data show that a substantial amount of vitamin C was excreted in the urine after the oral intake of vitamin C. These classical guidelines for the optimal dose of vitamin C should be reconsidered prospectively because vitamin C is a key substance for health promotion and disease prevention and not just for the prevention of scurvy. One more thing that should be considered to determine the RDA of vitamin C is that the beneficial effects of vitamin C on health promotion should be long acting. While preventing scurvy by taking vitamin C does not take a long time, the beneficial effects of vitamin C on health promotion take a longer time and sometimes lifelong in some cases. It means that it is actually very difficult to verify the beneficial effects of vitamin C on health promotion only through experiments, even though they are human trials. What is an actually reasonable guideline for the optimal dose of vitamin C? Strictly speaking, vitamin C is not a vitamin any longer because every mammal except for the primates, which include humans, produce vitamin C in their body. Therefore, first, one should consider the amount of vitamin C mammals are producing daily during their normal lives to help determine the actual optimal dose. Considering that guinea pigs, which cannot synthesize vitamin C by themselves, have a theoretical daily requirement of vitamin C about 5 mg/100 g body weight but in reality 10 mg or more is necessary, it can be seen that it needs about 10 g per day if it is converted into human case (Sutherland and Festing 1987). Furthermore, it is known that more than 70 mg of vitamin C per kg of body weight is produced daily in living mammals such as mouse, rat, pig, and so on (Woo 1999). They are producing even more vitamin C in stressful states, such as hard exercise or long exposure to the cold, which is 2~3times as much as in the normal resting state. Normally, animals are producing more than 100 times as much as the current classical recommended dose for humans. When considering the fact that the primary role of vitamin C is to maintain life, the current daily optimal dose of 60 mg should be increased up to 100 times that at least because the ability of the human body to maintain life is much less than that of animals, which are producing a substantial amount of vitamin C by themselves.

There is one more important thing to consider for the determination of the optimal dose, which is to observe the excretion pattern of vitamin C after oral intake. For the transportation of vitamin C, there are two transporter proteins, SVCT1 and SVCT2. SVCT1 is mainly distributed over the upper jejunum, liver, and proximal convoluted tubule of the renal tubular system, while SVCT2 is widely distributed over nearly all organs, but its distribution pattern is very organ dependent. For example, SVCT2 is highly expressed in the brain and adrenal gland, but it is rarely expressed in the muscles. SVCT1 is for the bulk transportation of vitamin C, while the expression of SVCT2 depends upon how much those organs utilize vitamin C. Here, the reason why SVCT1 is densely expressed on the surface of epithelial cells of the proximal convoluted tubule of the renal tubular system should be scrutinized. The role of SVCT1 in the proximal tubule is the reabsorption of vitamin C already filtered through the glomerular filtration barrier (Rumsey and Levine 1998; Hediger 2002). It is known that urinary excretion of vitamin C started to be observed from vitamin C intakes between 60 and 100 mg/day and that more than half of the vitamin C intake at 500 mg/day or more was excreted in the urine within 24 h (Graumlich et al. 1996; Levine et al. 1996a, 2001). Interestingly, all mammals except for primates normally produce vitamin C in their body, and their urine contains a large amount of vitamin C such that the urinary concentration (150–300 μM) reaches three to five times higher than that of the plasma level (our unpublished data). It is very unusual that any substance normally synthesized in mammals is filtered through the glomerular filtration barrier of the kidney. Furthermore, it is even more unusual that there should be an apparatus regulating the excretion of vitamin C by reabsorption in the epithelial surface of the proximal tubule (Fig. 1.5). These facts strongly suggest that the passing out of vitamin C through the kidney is not the excretion of a waste product or the overflow of an excess amount of some substances, and this excretion of vitamin C could have a very important biological significance. The original function of urination is the excretion of liquid waste products produced in the body after metabolism for maintaining life. Thus, actually urine also contains a substantial amount of oxygen free radicals, which are inevitably generated in the body during life processes, which subsequently cause oxidative damage to the surface of the epithelium of the urinary passage. Especially, the internal surface of the urinary bladder is likely to be damaged by the radicals most easily because it stores urine for a relatively long time especially during nighttime com-

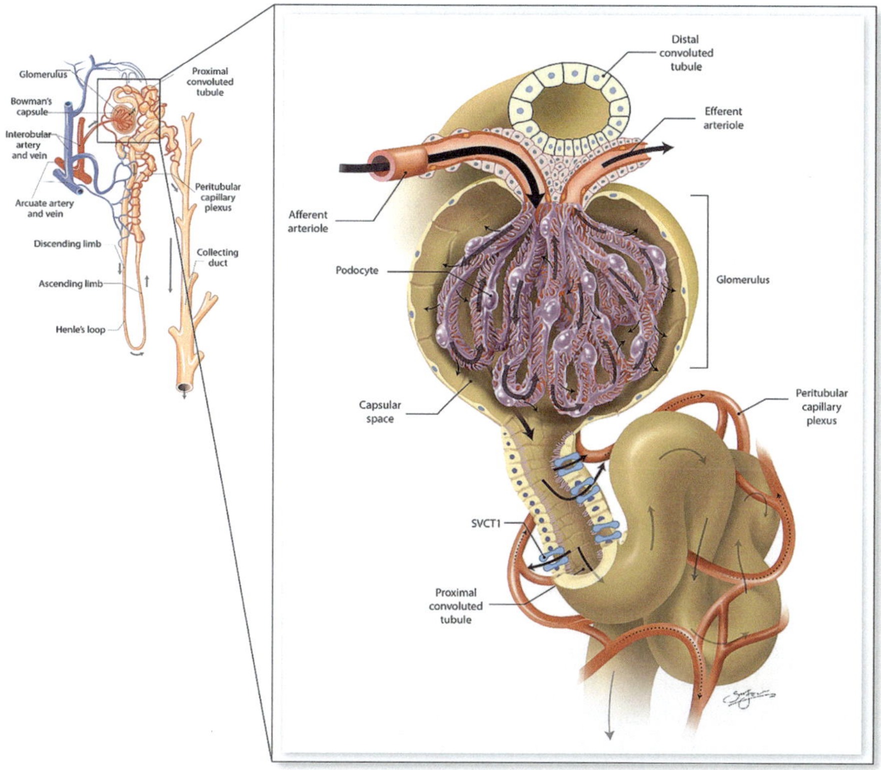

Fig. 1.5 Vitamin C excretion is controlled in the proximal tubule of kidney through SVCT-1. The vitamin C in the peripheral blood passes through the kidney as other molecules and then, some of it is excreted through glomerular filtration barrier. Interestingly, some of the filtered vitamin C is reabsorbed in the proximal tubule through SVCT-1, expressed heavily on the epithelium of the proximal tubule, and gets back to the blood via the perivascular capillary plexus. It is very uncommon that any substances, which are produced by themselves, are usually reabsorbed in the proximal tubule after filtration in the glomerulus. The arrows indicate direction of vitamin C flow and the thickness of arrows indicates relative amount of vitamin C

pared to daytime at which urination is more frequent. However, it seems there is no way to protect the surface epithelium lining the urinary bladder from oxidative damage even in many mammals synthesizing vitamin C by themselves. Passing out vitamin C through the kidney voluntarily, which is the representative antioxidant that can quench the radicals contained in urine, is the only way to protect the urinary bladder. That is the reason why even a very small amount of vitamin C is normally found in the urine after the oral intake of 60~100 mg of vitamin C. More interestingly, a relatively high amount of the pure form of vitamin C is normally contained in the urine of average mammals, which are producing vitamin C in their body. These facts, especially that vitamin C in the urine is not the excretion of waste materials but is for the protection of the urinary passage, should be considered in the determination of the optimal dose of vitamin C. Schlegel and his colleagues (1970) had already reported that the regression of bladder cancers in cigar and tobacco smokers could be possible by the oral intake of a relatively high dose of vitamin C (1 g or more daily). They found that the vitamin C levels in the urine of bladder tumor patients were low and those in cigarette smokers were half that of those in non-smokers. They also demonstrated experimental results using mice to elucidate a possible mechanism involved in the regression of tumors. They implanted a pellet containing 3-hydroxyanthranilic acid (a derivative of the amino acid, tryptophan) in a mouse bladder and found that bladder tumors were induced if the mice were fed a normal diet, but not if they were given extra vitamin C in their drinking water. Schlegel et al. suggested that vitamin C ingested by the mice might prevent the formation of carcinogens by inhibiting oxidation of 3-hydroxyanthranilic acid (Schlegel et al. 1970).

In these respects, one more thing to clinically speculate about the urinary damages caused by oxygen free radicals in the urine for the lifelong period is that many people usually experience frequent urination during the nighttime (nocturia) with age causing them to not fall into a deep sleep, frequently complaining of sleepiness, lethargy, and easy fatigue during daytime. Currently, causes of nocturia have never been known yet. Many urologists used to say that it is just an aging phenomenon. However, considering what is known to occur in the urinary bladder during the aging process, even though its pathogenesis should be proven through well-designed experiments, it can be speculated that senile nocturia would be induced gradually by chronic attacks of oxygen free radicals. In general, young people before 40 years of age do not experience nocturia, whereas after 50 years of age, its frequency increases with age.

One more scientific fact that should be considered to determine the optimal dose is that originally, average mammals including humans synthesized vitamin C in their liver, but only primates including humans have lost the ability to synthesize it in their liver. Biosynthesis of vitamin C starts with D-glucose, which is finally converted to L-gulono-γ-lactone by several enzymes. Oxidation of L-gulono-γ-lactone by some oxidases eventually leads to the biosynthesis of vitamin C in hepatocytes. The gene encoding L-gulono-γ-lactone oxidase is mutated in primates, resulting in the failure of vitamin C biosynthesis. The reason why and the time when primates failed to synthesize vitamin C in their body are not been known scientifically. Some investigators have explained the reason in terms of evolution. Namely, it has been thought that the dietary intake of food during the lifetime of humans contained a sufficient amount of vitamin C such that humans no longer needed to biosynthesize vitamin C, resulting in a mutation of the encoding gene. However, this premise is no more valid because there is no way to explain the reason why herbivore animals are still producing so much vitamin C even though the plants or fruits they ingest every day have a sufficient amount of vitamin C. This suggests that amount of vitamin C in the body that humans usually take is not sufficient for normal health promotion, and as such, a substantial amount of vitamin C equivalent to the amount of vitamin C produced daily by average mammals needs to be supplemented for human health promotion and disease prevention for a long life. Although some scientists (Higdon and Frei 2002) insist that 80% of neutrophils could be saturated with a daily intake of less than 100 mg of vitamin C, full saturation of neutrophils with vitamin C can be just an indicator for a diagnosis of the scurvy but can never guarantee health promotion and disease prevention for a long life.

Conclusively, the RDA of vitamin C should be raised up to a minimal amount of 6000 mg/day, which average animals are producing every day in their lives.

How to Take Vitamin C

So many people are taking vitamin C in excess, frequently up to several 100 times as much as the classical optimal dose, 60 mg/day. Actually, it is very easy to find a tablet containing 1000 mg of vitamin C on the market, which is more than 16 times higher than the classical optimal dose. Ironically, no guidelines for the administration of vitamin C have been documented based on scientific evidence. In my laboratory, a human trial for the correct intake of vitamin C was performed. Fourteen people (male, four; female, ten) participated. After an oral intake of 2–4 g of vitamin C during breakfast, 0.5 ml of peripheral blood was taken from each participant every hour for 12 h, and then, the blood concentrations of vitamin C were measured with HPLC. Overall, the concentration reached its peak at around 3 h after intake and came back down to the original level (before the oral intake of vitamin C) approximately 6 h after oral intake, suggesting that people ideally should take vitamin C every 3 h. Thus, they should take vitamin C six times a day at a minimum (Figs. 1.5 and 1.6).

However, it is actually impossible to do every day. A recent trend in the concept of drug administration is to reduce the number of administrations in a day to improve compliance. One time a day is the ideal drug administration. This trend shows that even if the vitamin C administration is not ideal, it is better to take vitamin C once a day rather than forgetting to take it throughout the day or eventually giving up because it is too burdensome to take it six times a day. Therefore, a compromise is to take vitamin C every 6 h, which is when people usually have their meals. In other words, people should take vitamin C during their meals.

In the stomach, food usually stays for 2–3 h during which some nitrous compounds have formed in an acidic environment, which are known to be possibly carcinogenic. Fortunately, it is known that vitamin C can prevent the formation of carcinogenic nitrous compounds (Mirvish 1994). These two facts should be considered when making a scientific guideline for the correct intake of vitamin C. Therefore, it is suggested that people should take vitamin C during every meal.

The last issue on how to take vitamin C is to determine the daily optimal dose. As discussed in the "Scientific Background on the Recommended Daily Allowance (RDA) for Vitamin C," the daily optimal dose should be elevated up to a minimal level of 6000 mg/day of vitamin C, which mammals normally produce every day.

Conclusively, it is recommended that people take 2000 mg of vitamin C three times a day, more specifically during every meal.

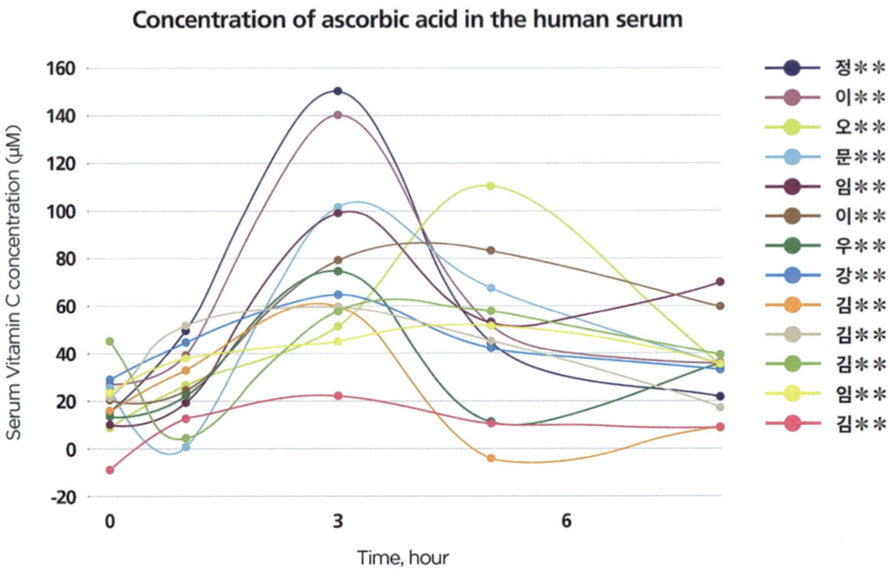

Fig. 1.6 Measurement of plasma vitamin C concentrations in healthy persons after oral administration of 2–4 g of vitamin C during breakfast. Most of the persons showed peak of blood concentration 3 h after administration and come back to the original basic level 6 h after oral administration, suggesting the reason why people should take vitamin C every 6 h (unpublished our data)

Adverse Effects of Vitamin C

Basically, vitamin C is a very safe chemical substance. Its adverse effects have started to be mentioned in earnest because many average people take much more vitamin C than the official RDA worldwide. The most common adverse effects mentioned are gastrointestinal (GI) problems because vitamin C is an acidic substance so that when administered in an empty stomach, the gastric mucosa may become irritated, even more severely in the case of gastritis or gastric ulcers. That is the reason why people should take vitamin C during meal times or just after a meal and not just before a meal. In reality, there are no gastric problems if people adhere to the rules of administration which is not to take vitamin C on an empty stomach. Therefore, GI problems as adverse effects are no longer a concern. Some people suffer from diarrhea only during the initial days of an administration of a relatively large amount of vitamin C. The diarrhea is likely occurring due to high osmotic challenge to the intestinal mucosa and does not last more than 1 week. Rather, it helps some people who have constipation. Therefore, it is also unlikely that diarrhea should be a serious adverse effect.

The only side effect to be concerned about is urinary stones. They are most commonly formed by the chemical reaction of calcium and oxalate during urine formation in the kidneys and urinary passage. There has been a report showing that high doses of supplemental vitamin C ranging 1000–10,000 mg/day increase oxalate excretion, while the official RDA of 20–60 mg/day does not lead to increased oxalate excretion (Johnston 2001). However, although the possibility of stone formation in the urine becomes higher due to the increase of oxalate in the peripheral blood, the increase of oxalate in the peripheral blood alone does not meet the condition for urinary stone formation fully. Nevertheless, in reality, according to my personal observation, when some people with an episode of urinary stone formation take supplemental high doses of vitamin C, the time interval of recurrence becomes shorter, indicating that taking high doses of supplemental vitamin C definitely promotes urinary stone formation in a subpopulation of people who meet the requirements for urinary stone formation. Despite this undeniable fact, competent urologists say that there is no way to prevent the formation of urinary stones except to drink a substantial amount of water every day. Most patients with an urinary stone usually suffer from colicky pain caused by stone attacks, radiating from the low abdomen to the groin region most likely during the hot summer season or after sweating heavily with vigorous exercise, which reduces the amount of water in the urine concentrating the oxalate and calcium so that the formation of an urinary stone is promoted and facilitated. In general, administration of high doses of vitamin C elevates the osmotic pressure of the plasma provoking thirst. Therefore, it is natural that people taking high doses of vitamin C every day drink much more water due to feeling thirsty than those not taking high doses of vitamin C. However, habitual neglect of such thirst in people's lives might lead to easy stone formation. Conclusively, if people with episodes of urinary stone formation want to keep taking supplemental high doses of vitamin C for their health, they must drink a substantial amount of water in their daily lives to reduce the number of urinary stone attacks.

There is one more obvious evidence that vitamin C is an extremely safe substance, even very high supplemental doses. Many healthy people are not rarely administered with an exceptionally high dose of vitamin C intravenously when they catch a severe cold or suffer from herpetic neuralgia (herpes zoster) or intractable fatigue. The usual range of doses of intravenous vitamin C is 10~50 g in an inoculation. Specifically, about 10–20 g of vitamin C are mixed in 500 ml of normal saline or dextrose saline and then dripped intravenously drop by drop for 3–4 h. Many cancer patients have been treated with high doses of vitamin C intravenously in many clinics. Briefly, their initial dose of intravenous vitamin C is 10 g a day, three times a week, and then, their doses are gradually increased to 30, 50, 70, and up to more than 100 g a day. In April 2017, the oncology team at the Iowa Medical Center reported that pharmacological ascorbate (an extremely high dose of vitamin C) administered intravenously with conventional anticancer treatment greatly improved the survival rate of glioblastoma multiforme (GBM) and non-small cell lung cancer (NSCLC) compared to the survival rate of the conventional treatment only. Even they administered vitamin C up to 125 g a day intravenously without any adverse effects (Schoenfeld et al. 2017).

References

Birlouez-Aragon L, Fieux B, Potier de Courcy G, Hercberg C. Vitamin C. In: Martin A, editor. Apports Nutritionnels Conseilles pour la Population Francaise. 3rd ed. London: Lavoisier Tec and Doc; 2000. p. 215–30.

Bjelakovic G, Nikolova D, Gluud LL, Simonetti RG, Gluud C. Antioxidant supplements for prevention of mortality in healthy participants and patients with various diseases. Cochrane Database Syst Rev. 2008;16(2):CD007176.

Bjelakovic G, Nikolova D, Gluud LL, Simonetti RG, Gluud C. Mortality in randomized trials of antioxidant supplements for primary and secondary prevention: systemic review and meta-analysis. JAMA. 2007;297(8):842–57.

Bruick RK, McKnight SL. A conserved family of Prolyl-4-hydroxylases that modify HIF. Science. 2001;294:1337–40.

Burri BJ, Jacob RA. Human metabolism and the requirement for Vitamin C. In: Packer L, Fuchs J, editors. Vitamin C in health and disease. New York: Marcel Dekker; 1997. p. 341–66.

Duffy SJ, Gokce N, Holbrook M, Huang N, Frei B, Keaney JF Jr, Vita JA. Treatment of hypertension with ascorbic acid. Lancet. 1999;354:2048–9.

Family of Mammalian Na+ dependent L-ascorbic Acid Transporters. Nature. 1999;399:70–5.

Food and Nutrition Board, Panel on Dietary Antioxidants and Related Compounds. Vitamin C, in dietary reference intakes for vitamin C, vitamin E, selenium, and carotenoids. Washington, DC: National Academy Press; 2000. p. 95–185.

Food and Nutrition Board. Recommended dietary allowances. 10th ed. Washington, DC: National Academy of Sciences; 1989.

Food and Nutrition Board. Vitamin C (Ascorbic Acid). In: Committee on Dietary Allowances, editor. Recommended dietary allowances. Washington, DC: National Academy of Sciences; 1980. p. 72–81.

German Society of Nutrition, Austrian Society of Nutrition, Swiss Society of Nutrition Research, Swiss Union of Nutrition. Recommended values for nutrient intake, 37–144, Umschau-Braus, Frankfurt, 2000.

Gewaltig MT, Kojda G. Vasoprotection by nitric oxide: mechanisms and therapeutic potential. Cardiovasc Res. 2002;55:250–60.

Graumlich JF, Ludden TM, Conry-Cantilena C, Cantilena LR Jr, Wang Y, Levine M. Pharmacokinetic model of ascorbic acid in healthy male volunteers during depletion and repletion. Pharm Res. 1997;14:1133–9.

Graumlich JF, King J, Cantilena LR. Vitamin C pharmaco-kinetics in healthy volunteers: evidence for a recommended dietary allowance. Proc Natl Acad Sci USA. 1996;93:3704–9.

Health and Nutrition Information Society. Recommended dietary allowances (dietary reference intakes). 6th ed. Tokyo: Daiichi Shuppan; 2000.

Hediger MA. New view at C. Nat Med. 2002;8:445–6.

Higdon JV, Frei B. Vitamin C: an introduction. In: Packer L, Traber MG, Kraemer K, Frei B, editors. The antioxidants C and E. 1st ed. California: AOCS Press, Santa Barbara; 2002. p. p1.

Huang A, Vita JA, Venema RC, Keaney JF Jr. Ascorbic acid enhances endothelial nitric-oxide synthase activity by increasing intracellular tetrahydrobiopterin. J Biol Chem. 2000;275:17399–406.

Jaakkola P, Mole DR, Tian Y-M, Wilson MI, Gielbert J, Gaskell SJ, Kriegsheim A v, Hebestreit HF, Mukherji M, Schofield CJ, Maxwell PH, Pugh CW, Ratcliffe PJ. Targeting of HIF-alpha to the von Hippel-Lindau Ubiquitylation complex by O2-regulated prolyl hydroxylation. Science. 2001;292(5516):468–72.

Jacob RA, Vitamin C. In: Shils ME, Olson JA, Shike M, Ross AC, editors. Nutrition in health and disease. 9th ed. Baltimore: Williams & Wilkins; 1999. p. 467–83.

Johnston CS. Vitamin C. In: Bowman BA, Russel RM, editors. Present knowledge in nutrition. 8th ed. Washington, DC: ILSI Press; 2001. p. 175–83.

Kuiper C, Dachs GU, Currie MJ, Vissers MC. Intracellular ascorbate enhances hypoxia-inducible factor(HIF)-hydroxylase activity and preferentially suppresses the HIF-1 transcription response. Free Radic Biol Med. 2014;69:308–17.

Levine M, Conry-Cantilena WY, Welch RW, Washko PW, Dhariwal KR, Park JB, Lazarev A, Graumlich JF, King J, Cantilena LR. Vitamin C pharmacokinetics in healthy volunteers: evidence for a recommended dietary allowance. Proc Natl Acad Sci USA. 1996a;93:3704–9.

Levine M, Rumsey S, Wang Y, Park J, Kwon O, Xu W, Amano N. Vitamin C. In: Ziegler EE, Filer LJ, editors. Present knowledge in nutrition. 7th ed. Washington, DC: ILSI Press; 1996b. p. 146–59.

Levine M, Wang Y, Padayatti SJ, Morrow J. A new recommended dietary allowance of Vitamin C for healthy young women. Proc Natl Acad Sci USA. 2001;98:9842–6.

Levine M. New concepts in the biology and biochemistry of ascorbic acid. N Engl J Med. 1986;314:892–902.

Li SH, Ryu JH, Park SE, Cho YS, Park JW, Lee WJ, Chun YS. Vitamin C supplementation prevents testosterone-induced hyperplasia of rat prostate by down-regulating HIF-1α. J Nutrit Biochem. 2010;21(9):801–8.

Malo C, Wilson JX. Glucose modulates vitamin C transport in adult human small intestinal brush border membrane vesicles. J Nutr. 2000;130:63–9.

May JM, Qu ZC, Nazarewickz R, Dikalov S. Ascorbic acid efficiently enhances neuronal synthesis of norepinephrine from dopamine. Brain Res Bull. 2013;90:35–42.

Mirvish SS. Experimental evidence for inhibition of N-Nitroso compound formation as a factor in the negative correlation between vitamin C consumption and the incidence of certain cancers. Cancer Res. 1994;54:1948S–51S.

Padayatty SJ, Sun H, Wang Y, Riordan HD, Hewitt SM, Katz A, Wesley RA, Levine M. Vitamin C pharmacokinetics: implications for oral and intravenous use. Ann Intern Med. 2004;140:533.

Page EL, Chan DA, Giaccia AJ, Levine M, Richard DE. Hypoxia- inducible factor-1alpha stabilization in nonhypoxic conditions: role of oxidation and intracellular ascorbate depletion. Mol Biol Cell. 2008;19(1):86–94.

Pauling L. Evolution and the need for ascorbic acid. Proc Natl Acad Sci USA. 1970;67:1643–8.

Rumsey SC, Levine M. Absorption, transport, and disposition of ascorbic acid in humans. J Nutr Biochem. 1998;9:116–30.

Schlegel IU, Pipkin GE, Nishmura R, Schultz GN. The role of ascorbic acid in the prevention of bladder tumor formation. J Urol. 1970;103:155–9.

Schoenfeld JD, Sibenaller ZA, Mapuskar KA, Wagner BA, Cramer-Morales KL, Furgan M, Sandhu S, Carlisle TL, Smith MC, Hejleh TA, Berg DJ, Zhang J, Keech J, Parekh KR, Bhatia S, Monga V, Bodeker KL, Ahmann L, Vollstedt S, Brown H, Shanahan Kauffman EP, Schall ME, Hohl RJ, Clamon GH, Greenlee JD, Howard MA, Shultz MK, Smith BJ, Riley DP, Domann FE, Cullen JJ, Buettner GR, Buatti JM, Spitz DR, Allen BG. $O_{2\cdot-}$ and $H_2O_2^-$ mediated disruption of Fe metabolism causes the differential susceptibility of NSCLC and GBM cancer cells to pharmacological ascorbate. Cancer Cell. 2017;31:1–14.

Sutherland SD, Festing MFW. The guinea pig in Chapter 24 of The Universities Federation for Animal Welfare. In Poole T, editor. Handbook on the care & management of laboratory animals, 6th ed, Published by Longman Scientific & Technical, 1987. p. 393–410.

Vissers MCM, Wilkie RP. Ascorbate deficiency results in impaired neutrophil apoptosis and clearance and is associated with up-regulation of hypoxia-inducible factor-1α. J Leukoc Biol. 2007;81(5):1236–44.

Woo HO. Vitamin C concentration in whole blood and plasma among residents living in southern central part of Korea. Dissertation of Gyeongsang National University College of Medicine, 2, 1999.

Yates AA, Schlicker SA, Suitor CW. Dietary reference intakes: the new basis for recommendation for calcium and related nutrients, B vitamins, and choline. J Am Diet Assoc. 1998;98:699–706.

Chapter 2
Cardiovascular System

It seems that most countries that have achieved successful economical and industrial development and cope with problems caused by a national food deficiency have experienced a sudden change in incidence pattern of disease. Especially, there is a trend for the rapid increase of the incidence of cardiovascular diseases. In general, the reason why the number of patients with cardiovascular diseases has rapidly increased can be explained in terms of a surplus energy intake. By examining the population pattern of disease occurrence more specifically, it can be seen that the age of disease initiation is getting younger and the incidence is drastically increasing. Among the mid-40 age group, vascular disease is prevalent, and even in the mid-30 age group, the disease incidence is reported quite often. Therefore, cardiovascular diseases are no longer limited only to the elderly. In addition, most of the increased cardiovascular diseases is mainly atherosclerotic vascular diseases. Furthermore, it has been reported that cardiovascular diseases have been the leading cause of mortality in the USA for the past 60 years, for which there is a clear relationship between the increased incidence rates of cardiovascular diseases and a surplus energy intake (Peters et al. 1998).

Cardiovascular System

This system, closely related with the maintenance of life, consists of the heart and blood vessels (arteries and veins). The heart acts as a pump that delivers two key substances for life, glucose and oxygen, to the cells comprising the whole body, which is called systemic circulation. One of the key substances for life, oxygen, is supplied by pulmonary circulation, which starts by passing the blood returned from the body to the right heart (right atrium and ventricle sequentially) to the lungs through the pulmonary arteries. After gas exchange of oxygen and carbon dioxide between the alveolar air space and capillaries in the alveolar walls of the lung, the blood rich in oxygen comes back to the left heart through the pulmonary veins

W. J. Lee, *Vitamin C in Human Health and Disease*,
https://doi.org/10.1007/978-94-024-1713-5_2

connecting with the systemic circulation. On the other hand, another key substance for life, carbohydrates, which react with oxygen producing energy to maintain life, are obtained by hepatic portal circulation, in which the venous tributaries drained from the GI tracts after arterial supply to them get together to form a large vein, the hepatic portal vein, which contains every nutrient absorbed through the GI tract. The hepatic portal vein goes into the liver to form an extensive capillary network over the whole area of the liver enabling the absorbed nutrients to be stored in the liver. It is connected to the hepatic vein after the extensive capillary network and finally to the right heart through the inferior vena cava. This is the hepatic portal circulation.

Heart

The heart is a fist-sized muscular organ. It is located in the middle of the chest cavity surrounded by both lungs and is one of the midline structures of the body with the apex elongated toward the left side and in the downward direction. The wall of the heart consists of three layers: inner layer (endocardium), middle layer (myocardium), and outer layer (epicardium). It has four chambers: the left and right atria and the left and right ventricles. There are valves present between the atria and ventricles; there is a tricuspid valve between the right atrium and right ventricle and a mitral valve between the left atrium and left ventricle. Two more valves are located at the outlet junction of the left (aortic valve) and right ventricle (pulmonary valve). The heart valves are check valves that allow blood flow in only one direction. The heartbeat is initiated in a small area (sinoatrial node; pacemaker) of the right atrium near the junction between the right atrium and the superior vena cava and is propagated to the atrioventricular node located at the bottom of the right atrium reaching the upper portion of the interventricular septum and is sent to the heart muscles of the left and right ventricles through the left and right bundle branches, respectively. Surprisingly, the heartbeat is initiated at a very early time of gestation maybe during the 4th week after fertilization, when the link between each organ and nervous system is not established yet. It means that the generation of the heartbeat itself is independent of nerve innervation. However, the quality of the heartbeat such as the strength or frequency of the beat is perfectly regulated by the autonomic nervous system as well as by some hormones (epinephrine, norepinephrine, and acetylcholine). This phenomenon is called “cardiac autonomy,” which is the anatomical basis for establishing the concept of brain death. The core concept of brain death is determined by the functional competence of the brain stem, of which the lower portion contains the respiration and heartbeat regulation centers. Fatal damage to the brain stem leads to an impairment of functions in both centers, resulting in the cessation of respiration and loss of control of the heartbeat with maintenance of monotonous feeble heartbeats followed by the natural course of death. However, if an artificial respirator can be provided to the patient, the patient can maintain a feeble heartbeat without consciousness called a brain-dead patient. A patient in a vegetative state is totally different from a brain-dead patient because the former maintains self-respiration in spite of loss of consciousness, while the latter cannot survive without an artificial respirator and is not aware.

All blood returning from the whole body to the heart goes into the right atrium through the superior vena cava from the head and neck areas and through inferior vena cava from the areas below the heart. The blood in the right atrium passes through the tricuspid valve to get into the right ventricle, where the blood is ejaculated into the pulmonary arteries, which distribute the blood evenly all over the lung tissue where the very crucial gas exchange (respiration) takes place. Blood oxygenated in the lung returns to the left atrium through the pulmonary veins and then flows into the left ventricle where the blood is strongly pumped out into the biggest artery, the aorta, which conducts all blood to the peripheral tissues (Fig. 2.1).

The heart is a muscular organ, and as such, it needs much blood. The heart is supplied with blood by the coronary artery, which is a very small artery and initiates at the junction between the cusps of the aortic valve and aortic wall. Ironically, blood is conveyed into the coronary arteries during the diastolic phase of the heart, while in general, blood is conducted into the arteries during the systolic phase of heart. There are left and right coronary arteries which have several branches to cover the entire area of the heart. The cardiac veins are accompanied by equivalent coronary arteries, and finally, the blood in those veins is returned directly into or by the coronary sinus into the right atrium (Fig. 2.2).

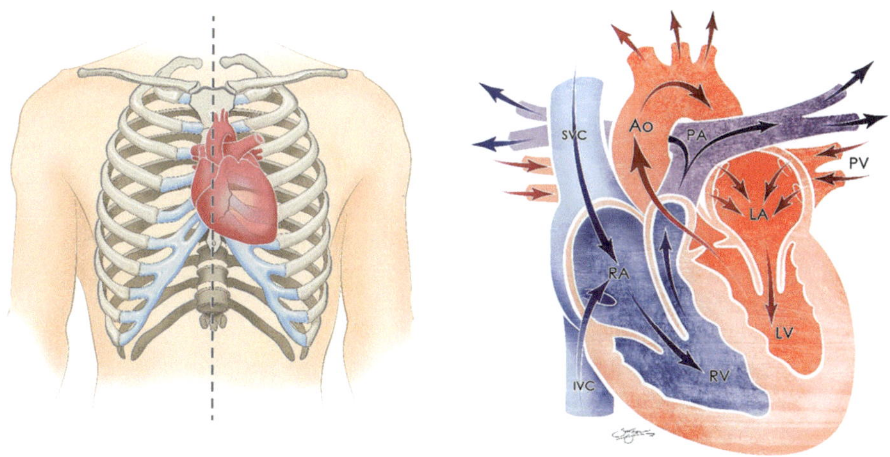

Fig. 2.1 Real position of the heart (left) and blood flows in the heart (right). Left figure shows that the heart is located at midline of the chest cavity. Right figure shows positions of each valve and directions of blood flow in the heart. Tricuspid valve (T) is located between RA and RV; mitral valve (M) is located between LA and LV. Aortic and pulmonary valves are located at roots of aorta and pulmonary artery, respectively. *SVC* superior vena cava, *IVC* inferior vena cava, *RA* right atrium, *RV* right ventricle, *PA* pulmonary artery, *PV* pulmonary vein, *LA* left atrium, *LV* left ventricle, *Ao* aorta

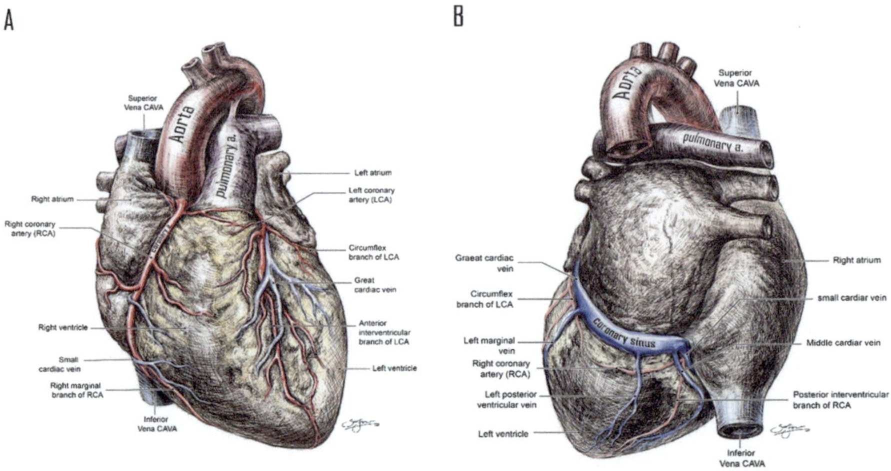

Fig. 2.2 Coronary arteries (red) and cardiac veins (blue). (**a**) Front view. The right coronary artery is initiated at right anterior side of the aortic root, while left coronary artery is initiated at left posterior side of the aortic root, and then divided into circumflex branch backward and anterior interventricular branch forward. The great cardiac vein is drained into coronary sinus, and the small cardiac veins are drained into the right atrium directly. (**b**) From behind. The circumflex branch of LCA is connected to the terminal branch of RCA, from which the posterior interventricular branch is directed downward and connected to the terminal branch of the anterior interventricular branch of LCA. The left marginal vein, the left posterior ventricular vein, and the middle cardiac vein are drained into the coronary sinus, which is finally drained into the right atrium. *RCA* right coronary artery, *LCA* left coronary artery

Arteries and Veins

The vessels containing the blood going from the heart to the cells, the final destinations of the cardiac circulation, are, by definition, the arteries. In general, there are numerous arterial branches, and the caliber of the arteries gets smaller in the direction of the blood flows, and the arteries terminate as a capillary network, which consists of vessels with the smallest caliber that can directly supply the cells with the blood. Meanwhile, the vessels containing the blood returning to heart after arterial supply are, by definition, veins. Therefore, many tributaries of vessels containing returning blood run together to form small-sized veins, medium-sized veins, and, finally, large veins such as the inferior and superior vena cava connected to the right heart. Surprisingly, the total length of the vessels has been known to be tens of thousands of miles (two times as long as the circumference of the earth), most of which is occupied by the microvascular network, which consists of small arteries (arterioles), capillaries, and small veins (venules).

The structures of the arteries and veins commonly have three layers, the intimal layer (inner layer), middle layer (media), and adventitia layer (outer layer), which are quite different from each other. Generally, the intimal layer functions just as an inner boundary and covering such that in some small vessels, it consists of just one endothelial layer, while the adventitial layer, which consists of connective tissue fibers, functions as the outer boundary and covering. However, the middle layer usually has a key role in performing the vessel functions. For example, the main function of arteries is to conduct the blood sufficiently and efficiently to the cells, and this function is performed primarily by the middle layer. Actually, the importance of the middle layer is reflected by its structure, which is usually filled with elastic fibers and smooth muscles. Especially in arteries, the middle layer has the pivotal role in conducting and generating the blood pressure. In large arteries such as the aorta and iliac arteries, which do not have to generate the blood pressure but conduct the blood that is strongly pumped out from the left ventricle, their middle layer has half the elastic fibers and smooth muscles compared to small arteries. They have to generate the blood pressure because the blood pressure generated and conveyed from the heart becomes extinct by intimal resistance along the circulatory path. Thus, smooth muscles, which enable arteries to contract voluntarily, are dominate in the middle layer of small arteries. Veins have completely different features. Their outer layer among three layers is prominent and rich in interwoven connective tissue fibers (Fig. 2.3).

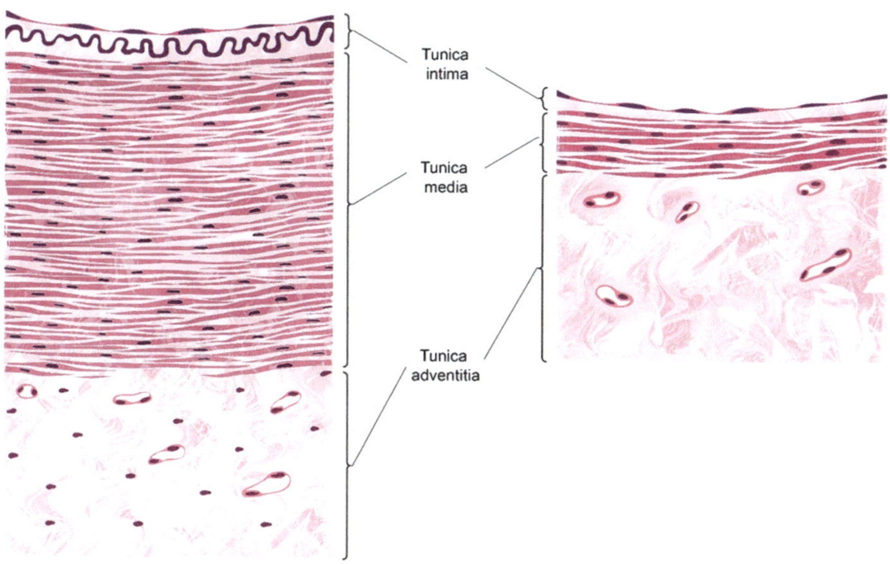

Fig. 2.3 Structures of artery and vein. Artery has prominent middle layer (media), while vein has the thickest outer layer (adventitia) as well as intima consisting of a single endothelial layer. Total thickness of arterial wall is usually 2–3 times as thick as venous one

Blood Pressure

Laypeople usually say that normal blood pressure is 120/80, which means that the systolic pressure is 120 and the diastolic pressure is 80. Specifically, 120 means that the pressure generated in the arterial system during the heart contraction is equivalent to the power elevating the mercury (Hg) to 120 mm of height. Because the specific gravity of mercury is 13.6, 120 mmHg is the same as 163 cmH_2O; in other words, normal cardiac pumping can expel blood to more than 1.6 m of height. The blood pressure generated in the venous system reaches less than 3 cmH_2O, while that in the arterial system reaches even more than 3 mH_2O. It is, therefore, amazing that the blood reaching the bottom of the foot can be returned to the heart against gravity in a moment. There are a couple of driving forces enabling successful venous return to the heart, even though the venous pressure can only elevate the blood just 2–3 cm in height. The most important driving force is capillary action, which partially enables the venous blood to be elevated against gravity. Second, the venous system located below the heart is equipped with check valves at 2–3-cm intervals, while that above the heart is not. Lastly, respiration has a minimal effect on successful venous return of blood. The negative pressure that occurs in the chest cavity during expiration can only help slightly elevating the blood from below.

Atherosclerotic Vascular Diseases

Atherosclerotic vascular diseases are diseases confined to all the various kinds of arteries. Clinically, there are two kinds of diseases: arterial obstruction and bleeding. And even though atherosclerotic diseases may occur in any organ of the body, there are four major organs causing serious clinical problems due to atherosclerotic diseases: the brain, heart, kidney, and retina. The diseases in the brain are cerebral infarctions (obstruction) and cerebral hemorrhages (bleeding), while that in the heart is only myocardial infarction (obstruction). Atherosclerotic disease in the kidney usually results in chronic renal failure, for which hemodialysis is inevitable for life. Atherosclerotic retinal disease usually causes people to suffer from visual problems, sometimes even blindness. Miscellaneously, some people suffer from weakness or tingling sensation due to atherosclerotic vascular diseases occurring in the upper and lower extremities. Fortunately, the pathogenesis of atherosclerosis is well documented.

Pathogenesis of Atherosclerotic Disease

According to the dictionary's definition, atherosclerosis is an extremely common form of arteriosclerosis in which deposits of yellowish plaques (atheromas) containing cholesterol, lipoid material, and lipophages are formed within the intima and inner media of large- and medium-sized arteries. There are two clearly defined conditions for the formation of atherosclerotic lesions in the arterial endothelium. The first condition is micro-damage to the arterial endothelium. The second one is the peroxidation of cholesterol, especially low-density lipoproteins (LDLs), which are conjugates of carrier proteins and cholesterol. Oxidized LDLs are known to contribute to the development of atherosclerosis (Luc and Fruchart 1991; Abbey et al. 1993). Once the arterial endothelium is injured by any cause, foamy cells, which are known to be monocytes engulfing oxidized LDLs, are recruited to the damaged site of the arterial endothelium forming fatty streaks and atherosclerotic plaques. Atherosclerotic lesions have never been found in any arterial endothelium that has not experienced micro-injury. This is the reason why atherosclerotic vascular diseases have never been found in young age groups without any exceptions.

Classically in medicine, hypertension and diabetes mellitus (DM) are known to be the major causes of atherosclerotic vascular diseases. The prolonged exposure of the arterial endothelium to high arterial blood pressure over 130 mmHg is bound to induce micro-damage to the arterial endothelium. Especially, the damage is known to occur preferentially at the site of the arterial bifurcation by inducing a turbulent blood flow. Actually, the preferred sites of atherosclerotic arterial lesions induced by hypertension are the bifurcating points of the arteries. Interestingly, British medical scientists, Bates and his colleagues, reported that in an epidemiological study, in which more than 500 men and women participated, vitamin C levels in the peripheral blood were inversely correlated with systolic and diastolic blood pressure (Bates et al. 1998). Therefore, the importance of keeping the arterial systolic blood pressure under 130 mmHg cannot be overemphasized. The case is exactly the same in diabetes mellitus (DM). Prolonged exposure of the arterial endothelium to high blood sugar over 120 mg/dl 2 h after a meal is likely to start endothelial damage. It has been known that the diabetic patients show a condition of elevated oxidative stress (Devaraj et al. 2001; Samson et al. 2002). Furthermore, according to some observational studies, the blood levels of vitamin C in the diabetic patients were lower than those of nondiabetic healthy people by at least 30% (Will and Byers 1996). In addition, a cross-sectional study using more than 6000 British men and women told us that the blood vitamin C levels were inversely correlated with glycosylated hemoglobin (HbAlc) (Sargeant et al. 2000).

Nowadays, among relatively young age groups without hypertension or DM, atherosclerotic vascular patients have progressively increased. Interestingly, even though some patients have never experienced hypertension or high blood sugar, they have suffered a fatal atherosclerotic arterial obstruction to the extent that they had to be rescued by an interventional treatment modality, such as a stent. What has a critical role in the induction of arterial endothelial damage besides hypertension and high blood sugar? The significance of reactive oxygen species (ROS) cannot be overlooked. ROS is generated during normal human metabolic processes for maintaining life. ROS generation should never be a pathological phenomenon and only a physiological one. Although it stimulates normal physiological processes, it is well known that it can induce chronic endothelial damage over one's life. Many papers have reported that ROS can induce endothelial damage (Paduch et al. 2005; Nakagiri et al. 2007; Schepers et al. 2009; Pasciu et al. 2010; Rovira-Llopis et al. 2013;

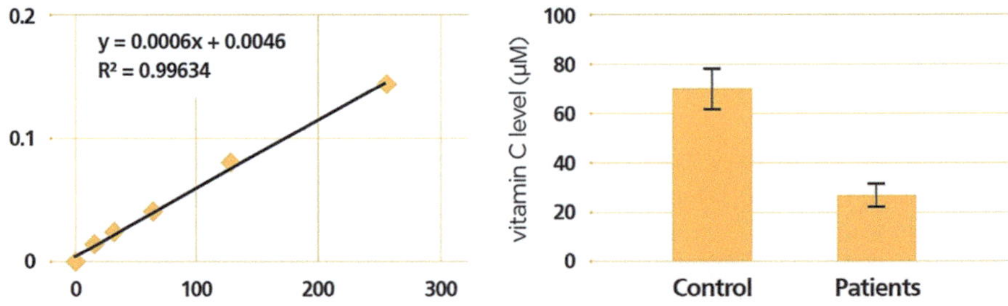

Fig. 2.4 Plasma vitamin C level in myocardial infarction patients was severely reduced, comparing to that in healthy control (Unpublished our data)

Lubrano and Balzan 2014). Interestingly, our unpublished data show that the plasma vitamin C levels in patients with myocardial infarction were much lower than in the healthy controls (Fig. 2.4).

It has been well documented that the concentration of several antioxidants including vitamins C and E is decreased in the peripheral blood of DM patients compared to normal healthy people (Fain et al. 2003), which strongly suggests that high blood sugar might induce a condition in which more substances causing oxidative damage are generated. Especially, the generation of oxygen free radicals has increased in the lives of modern people for whom overeating and extreme stress are inevitable. Extreme stress necessarily accompanies marginal hypertension and high blood sugar for more energy production, which needs to cope with stress, followed by increased generation of oxygen free radicals. In classical medicine, the major two factors for atherosclerotic lesions are hypertension and DM. Additionally, there are two aggravating factors which are cigarette smoking and obesity. How can these two conditions accelerate the formation of atherosclerotic lesions? It has been well documented that there are a couple hundred substances causing oxidative stress in cigarette smoke. Actually, there are reports that cigarette smoking lowers the plasma vitamin C level (Pelletier 1968; Kaller et al. 1979). Furthermore, the vitamin C level in serum lowered by cigarette smoking may lead to an increased risk of atherosclerotic vascular disease by oxygen-derived free radicals (Machlin and Bendich 1987; Jacob 1994). Obesity makes the body need more energy; thus, it is quite natural that more oxygen free radicals are generated. Therefore, there is no doubt that cigarette smoking and obesity can aggravate atherosclerotic lesions induced by hypertension or high blood sugar.

Conclusively, it is quite obvious that modern people, to whom overeating and extreme stress are inevitable fate, cannot avoid early onset and high incidence of atherosclerotic vascular disease unless they have appropriate measures to control hypertension and high blood sugar and to quench increased ROS.

The next topic that should be considered for the formation of atherosclerotic vascular lesions is about cholesterol. Only cholesterol oxidized due to oxidative damage in the peripheral blood can be engulfed by monocytes on the membrane for which the expression of a surface marker for oxidized cholesterol is induced. The monocytes engulfing the oxidized cholesterol are called "foamy cells," which are recruited to the injured site of the arterial endothelium forming the atheroma plaques. Needless to say, the main cause for the peroxidation of cholesterol is the action of ROS. Cholesterol is normally synthesized in the liver and fat cells. It has been known that about 70–75% of total cholesterol is synthesized in the liver and fat cells and that the remaining 25–30% is absorbed from the daily diet. Cholesterol has several important functions in the body such as a key component of bile juice, a biochemical precursor of steroid hormone, a component of cell membrane, a role in the formation of myelin sheath, and so on. For its transportation, it is usually conjugated with carrier proteins, and this complex is called a lipoprotein. Initially, in the stage of hepatic biosynthesis, a greater amount of cholesterol is conjugated with carrier proteins, and those complexes are called very-low-density lipoproteins (VLDLs), which are sustained for a very short period. On their way to delivering cholesterol to the target organs, the proportion of cholesterol decreases so that the density of the complex becomes a little bit increased forming low-density lipoproteins (LDLs), which have a wide range of density and occupy the greatest proportion among the total cholesterol. Finally, used cholesterol in the cells is usually recycled in the form of high-density lipoproteins (HDLs). It is obvious why LDL cholesterols have been called bad cholesterol which has a key role in the formation of atherosclerotic vascular change. LDL cholesterols have the greatest proportion of cholesterol compared to HDL cholesterols, so that the peroxidation of cholesterol more easily occurs with LDL cholesterols than with HDL cholesterols. Once LDL cholesterols are oxidized, they are more readily taken up by macrophages compared to non-oxidized LDLs, which is the conversion process of macrophages to foamy cells. Aggregation of these cells progresses to form fatty streaks and plaques later. Oxidized LDLs inhibit the migratory function of macrophages which subsequently results in the macrophages sitting on the intima layer and promotes the uptake of oxidized LDLs. They are also so cytotoxic that they usually induce cell death, endothelial loss, and denudation of the artery. Some antioxidants such as vitamins C and E can prevent the oxidation chain reaction-enabled LDL cholesterols oxidized by quenching ROS through hydrogen donation (Abbey et al. 1994). Fortunately, only vitamin C is reported to prevent the initiation of the oxidation chain reaction in lipids.

Because plasma antioxidant levels such as vitamins C and E have been known to be closely related to the inflammation process (Bayata and Lau 2000; Beharka et al. 2000; Bowie and O'Neil 2000) and vitamin C has also been known to have a uricosuric effect (Stein et al. 1976; Berger et al. 1977; Ersoy et al. 1999), the relationship of the serum vitamin C level to some nontraditional atherosclerotic vascular disease risk factors involved in the inflammation process, such as C-reactive proteins, plasma fibrinogen, and leukocyte count, as well as some other factors, such as serum uric acid, homocysteine, creatinine, and albumin, was examined in the data from the NHANES III (7345 women and 6390 men). Interestingly, the levels of serum vitamin C were inversely correlated with the serum levels of creatinine and homocysteine in both sexes with high statistical significance (P values less than 0.02) after adjusting for potential confounders such as smoking and alcohol ingestion. However, there were no correlations between the serum vitamin C levels and the uric acid levels or the white blood cell counts in both sexes. While the serum vitamin C levels were inversely correlated with the plasma levels of fibrinogen in women, they were directly associated with the serum level of albumin in men (p value is 0.02) (Simon 2002).

Taking everything mentioned above together, the risk factors of atherosclerotic vascular disease can be categorized into two groups, traditional and nontraditional ones. The traditional risk factors are high blood cholesterol, hypertension, diabetes, obesity, and smoking. Unfortunately, these traditional risk factors cannot account for the pathogenesis of all cases of atherosclerotic vascular diseases. Rather, recently, nontraditional risk factors, which are all factors except for the traditional ones, are becoming more important because the number of patients without the traditional risk factors is increasing. Furthermore, among the many nontraditional risk factors, ROS seem to have a key role in the formation atherosclerotic vascular lesions because ROS can fulfill the two conditions (arterial endothelial damage and lipid peroxidation) for atherosclerotic vascular changes simultaneously. At the end of the twentieth century, Dr. Gey had already hypothesized that chronic oxidative insult could change LDL cholesterols resulting in vascular damage and eventually atherosclerosis (Gey 1986, 1994). He proved his hypothesis in part by showing that low plasma levels of antioxidants such as vitamins C and E may be modifiable risk factors for atherosclerotic vascular lesions (Gey 1998). Surprisingly, in 2009, Dr. Gey's hypothesis got strong support. In addition, there was a dramatic reversal in the medically orthodox concept of atherosclerotic vascular diseases. In reality, atherosclerotic vascular changes have been known for a long time in the clinical field to be irreversible processes. However, strikingly, it has been postulated that atherosclerotic lesions could be reversed by prolonged intake of strong antioxidants such as resveratrol which is present in red wine. Linda reported a mechanism that is possibly involved in the reversing process, which made sure that the known pathogenesis of atherosclerosis described above should be correct (Fig. 2.5) (Curtis 2009).

Conclusively, ROS have a pivotal role in both prerequisite conditions (the endothelial damage of arteries and peroxidation of cholesterol) for the formation of atherosclerotic lesions. Therefore, any efforts to reduce the amount of ROS generated in normal life should never be neglected, and people should do their best to keep their blood pressure and plasma sugar level within the normal limit for the health of the cardiovascular system. Needless to say, the smoking should be prohibited, and well-planned exercise program is recommended to keep body weight within the normal limit.

The formation of an atherosclerotic plaque depends on monocyte entry and CD36-dependent macrophage trapping in the arterial intima. A recent study by Park et al. showed that oxidized low-density lipoprotein (LDL) interacts with CD36 on the macrophage to facilitate cholesterol loading and to activate a signaling cascade that leads, through the generation of intracellular reactive oxygen species (ROS), to the loss of dynamic control of cell adhesion and thus an increase in actin polymer-

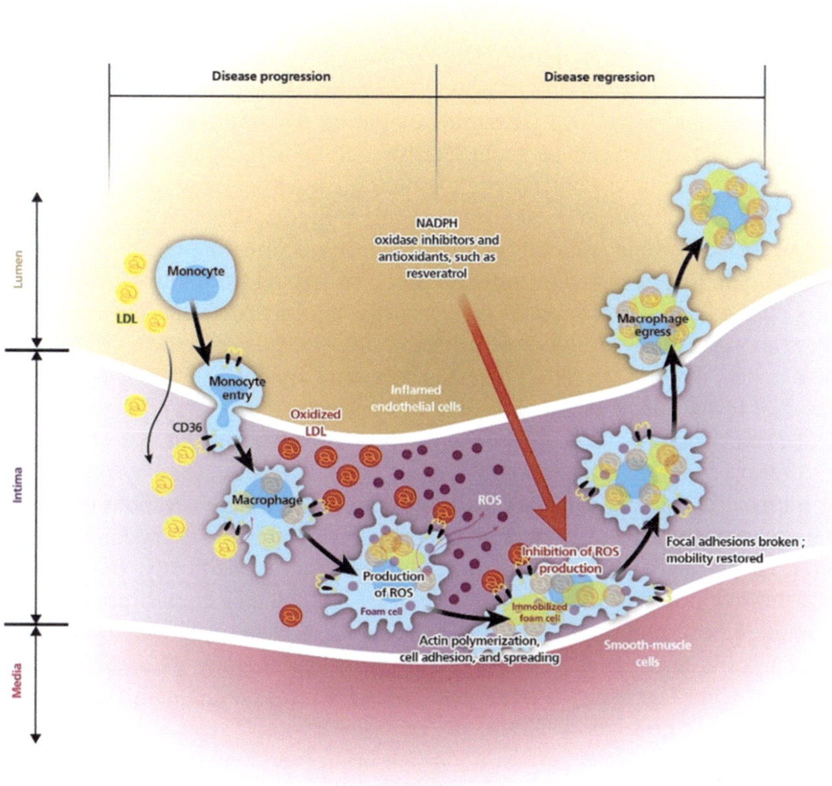

Fig. 2.5 Proposed mechanism of how the atherosclerotic vascular disease progresses and can regress. The atherosclerotic plaque is usually formed by monocyte entry and CD36-dependent macrophage trapping in the arterial intima. Interaction of oxidized low-density lipoprotein (LDL) with CD36 on the macrophage facilitates cholesterol loading and activates a signaling cascade leading to loss of dynamic control of cell adhesions through the generation of intracellular reactive oxygen species (ROS) and thus an increase in actin polymerization. Finally, cell spreading is enhanced with concomitant inhibition of migration in vitro. This cellular events explain processes of foam cell entrapment in the intima, a characteristic of disease progression. Fortunately, macrophage mobility can be restored with the use of antioxidants and inhibitors of NADPH oxidase, a finding consistent with the possibility of foam-cell egress. (Curtis 2009).

ization (Park et al. 2009). The net effects are the enhancement of cell spreading with concomitant inhibition of migration in vitro. This cellular behavior is consistent with the entrapment of foam cells in the intima, a characteristic of disease progression. Park et al. observed that macrophage mobility can be restored with the use of antioxidants and inhibitors of NADPH oxidase, a finding consistent with the possibility of foam-cell egress (Curtis 2009).

Vitamin C and Nitric Oxide (NO)

NO is a very important signaling molecule in the human body, and it is, especially, even more powerful and important in the cardiovascular and nervous system. NO has many functional effects on all organs including the stomach, kidneys, lungs, liver, genitals, and heart because it can send specific biological signals through the cell membrane, and those signals regulate biological activities as well as various functions. The main function of NO is vasodilation. Medical doctors specializing in the cardiovascular system sometimes prescribe nitroglycerine to patients complaining of chest pain that occurred during exercise. NO derived from nitroglycerin sublingually applied to patients induces the vasodilation of narrowed coronary arteries leading to enough supply of blood to the heart muscles resulting in the relief of chest pain also called "angina pectoris." Another important function of NO is to prevent intravascular formation of blood clots. Nowadays, more people are regularly taking low-dose aspirins to prevent intravascular clotting. Actually, the number of patients today suffering from vascular problems of the brain or heart caused by the formation of intravascular clots is increasing. For example, the varices (venous engorgement) frequently reported in outpatient clinics for vascular diseases are known to be the major cause of the formation of intravascular clots. Specifically, the slowing down of the blood flow due to venous varicosity is known to be a direct cause of intravascular clot formation. In addition, NO can also have a small role in lowering the plasma cholesterol level especially in combination therapy with statin drugs such as Lipitor and Mevacor which are known to increase and maintain NO production. Furthermore, the roles of NO in the central nervous system cannot be overlooked because it is used for several important brain functions such as the storage and retrieval of long-term memory as well as the transmission of incoming and outgoing neural information. Actually, it looks like there is no end to the uses of NO in the human body because it must be one of the powerful antioxidants.

Fortunately, the human body produces NO. It is synthesized mainly in the endothelium of the blood vessels that have a total length that reaches approximately 100,000 km, all of which must have the endothelium. The endothelium is a film-like thin tissue forming a demarcation of blood and tissues. Malnourishment or any kind of damage to it hampers the endothelium to produce a sufficient amount of NO. In fact, the endothelium can synthesize NO with the help of an enzyme called endothelial NO synthase (Ignarro 2000). Interestingly, vitamin C is a crucial cofactor for that enzyme, suggesting that vitamin C can have an important function in vasodilation, which is one of the key mechanisms involved in lowering high blood pressure, which must be one of the chronic etiologies for atherosclerotic vascular diseases. Actually in 1999, the medical researchers working in Boston University and Linus Pauling Institute of vitamin C had reported that vitamin C has a significant clinical effect on lowering high blood pressure in a couple hundred patients that visited the outpatient clinic using a randomized, placebo-controlled, double-blinded test (Duffy et al. 1999).

French Paradox

In early 1990, it was reported that the average French person has a relatively lower incidence of atherosclerotic coronary heart disease than the average European person, even though French people have a relatively higher level of cholesterol in the peripheral blood compared to European people. In terms of cholesterol, which is known as a main cause of atherosclerosis, ironically, the present medical orthodox concept on atherosclerotic vascular disease maintains firmly that people with a higher level of blood cholesterol level have a higher incidence of atherosclerotic vascular disease. If so, why has such a paradoxical phenomenon taken place only in people from some French areas? Among the notable differences in diet between these two people, the difference in the consumption amount of red wine is noteworthy. French people, which live in areas with many wineries such as Bordeaux or Bourgogne, are known to usually drink more red wine than the neighboring European people. This phenomenon suggests that red wine itself or some component of it might prevent or delay atherosclerotic vascular damage. Although this phenomenon is reported as an epidemiological study, it must have a very significant clinical reason that surely reinforces the scientific background on the pathogenesis of atherosclerosis, which tells us that the key point in prevention or reduction of atherosclerotic vascular damage is to minimize oxidative injuries as much as possible so that endothelial damage to arteries is prevented and the peroxidation of cholesterol is minimized, finally resulting in a decrease of atherosclerotic vascular disease. What component in red wine can have such a role? Resveratrol, which is known as a strong antioxidant, is usually only soluble in organic solvents. The alcohol in red wine is a kind of organic solvent which can extract resveratrol from the grape peel during fermentation. On the other hand, there is no resveratrol in grape juice because it is not soluble in water. This is the reason why grape juice has no preventive effects on atherosclerotic vascular diseases.

Scientific Background on the Potential Beneficial Effects of Vitamin C on the Prevention of Atherosclerotic Vascular Diseases

Vitamin C is known to have several preventive effects on the formation of cardiovascular disease. First, it can prevent LDL peroxidation by ROS, which is a very important prerequisite for the formation of atherosclerotic vascular lesions, and it can reduce endothelial damage to arteries, which is the starting point for the formation of atherosclerotic vascular lesions, caused by ROS by quenching them. Vitamin C is the most ideal biological antioxidant. In this regard, vitamin C can also regenerate oxidized vitamin E by reducing it back to its original form, vitamin E, which is a stronger antioxidant than vitamin C. Vitamin E has been reported to slow the rate of LDL oxidation (Abbey et al. 1993; Frei 1991). Therefore, if people take vitamin

E after they take a large amount of vitamin C, atherosclerotic vascular disease could be prevented more effectively. In addition, vitamin C is an essential cofactor for collagen synthesis (Jacob 1994), which helps in wound healing. Considering the composition of the vascular wall, it is also very important to maintain vascular integrity (Simon 1992). It has also been reported that Vitamin C may increase the plasma level of HDL cholesterol and decrease the plasma level of total cholesterol at least under certain conditions, indicating a marked decrease in LDL cholesterol, which is mainly related with the pathogenesis of atherosclerotic vascular diseases (Simon 1992; Simon et al. 1993; Jacques et al. 1994; Hallfrisch et al. 1994). In addition, vitamin C affects cholesterol metabolism so that it may stabilize the level of plasma cholesterol because it has some roles in enzyme production involved in the biosynthesis or hydroxylation of cholesterol (Carr and Frei 1999). This scientific background strongly supports that clinicians should recommend suitable antioxidants such as vitamin C or vitamin E or, even better, a combination of vitamins C and E to prevent LDL cholesterol from being oxidized rather than recommend taking medication to lower the plasma level of LDL cholesterol.

Many people are recommended by their physician to take low-dose aspirin every day to prevent intravascular thrombosis. Unfortunately, laypeople usually have a misconception about the effect of low-dose aspirin which is that it can strengthen their blood vessels. Instead, people who take low-dose aspirin every day are at risk for uncontrolled bleeding in any organ that was damaged by an accident or by a pathological condition such as cerebral stroke or rupture of arterial aneurysms. By the way, vitamin C also has weak antithrombotic and antiplatelet effects through its interaction with the arterial endothelium to synthesize prostacyclin (Simon et al. 1983; Srivastava 1985; Toivanen 1987; Lefer 1990). Therefore, if you already have started to take a sufficient amount of vitamin C, you do not have to take low-dose aspirin routinely every day for the prevention of intravascular clotting, which is a critical cause of some vascular incidents such as cerebral stroke in some cases. However, some patients such as those patients who were diagnosed with arrhythmia, patients who underwent surgery for valvular heart disease, patients treated with a stent due to the narrowing of the coronary arteries, or those who have venous varicosities in their lower legs should take low-dose aspirin prescribed by their physician.

Vitamin C can also promote nitric oxide production which must be helpful in maintaining a normal range of blood pressure in hypertension patients by its induction of vasodilation (Ting et al. 1997; Timini et al. 1998). Especially, some cardiologists working in Boston Medical Center performed a collaborative clinical study on the effects of vitamin C on hypertension with the medical scientists working at the Linus Pauling Institute at Oregon State University. They clearly showed that vitamin C could lower blood pressure successfully in a randomized, double-blinded, placebo-controlled study. Although, unfortunately, they could not find any evidence involved in the reduction mechanism of blood pressure, such as nitric oxide bioactivity or prostacyclin production, as well as evidence related to the pathogenesis of vascular disease such as reduced lipid peroxidation or production of F_2 isoprostanes, it is memorable that this must have been the first clinical report of vitamin C performed in a randomized, placebo-controlled, and double-blinded manner (Duffy et al. 1999). The proper control of hypertension by any means cannot be emphasized enough because lifelong hypertension eventually leads to atherosclerotic vascular diseases.

Taken all together, although there is no doubt that people should take a large amount of vitamin C with every meal for the prevention of atherosclerotic vascular lesions as well as for the maintenance of a healthy arterial condition, most of the epidemiological studies on the effects of vitamin C on atherosclerotic vascular diseases scarcely show positive results. What are the real causes of such discrepancies? Next, the overall trend of epidemiological studies on the relationship between vitamin C and atherosclerotic vascular diseases will be reviewed.

Review of Epidemiological Studies on Relationship Between Vitamin C and Atherosclerotic Vascular Diseases

Because vitamin C can exert its beneficial effects for a very long period, sometimes for a lifelong period, it is very hard to demonstrate its effects through experiments, which have clear limitations such as a short experimentation period, limited artificial conditions, or a limited number of animals. In addition, although it is closely related to human health, it is impossible to perform direct in vivo experiments using humans, and thus, it is inevitable to perform the experiments using animals. However, because most animals, especially mice, usually synthesize vitamin C themselves in their bodies, it is very hard to observe the authentic effects of exogenous vitamin C. In addition, once exogenous vitamin C is administered, it is quite natural that the experimental animals reduce the production of endogenous vitamin C. Therefore, in many of the cases of animal studies, positive results are not always reported. Fortunately, the gulo(−/−) mice, which cannot synthesize vitamin C by themselves because the gene encoding L-gulono-γ-lactone oxidase is knocked out just like in humans, were developed in 1999 by Dr. Hediger's group. Since then, many reliable positive results have been published through experiments using the gulo(−/−) mice (Jeong et al. 2011; Kim et al. 2012, 2013, 2014, 2015; Bae et al. 2013; Nagappan et al. 2013a, b; Yu et al. 2015). Despite the many publications on the positive results of the beneficial effects of vitamin C on human health obtained from animal studies, scientific pursuit strongly requires clinical research with humans showing the practical effects of vitamin C on the cardiovascular system, especially in the formation of atherosclerotic vascular lesions. It is not so easy to perform clinical research with humans and vitamin C. The first clue for the relation between vitamin C and coronary heart disease comes from James Lind's clinical observation of unexpected cardiac death among seamen with scurvy in 1757 (Lind 1757). Long later, although several clinical studies have been done by medical scientists inspired by Dr. James Lind's insight, they only did a simple comparison between consumers and nonconsumers of vitamin C, or most of the studies have been performed epidemiologically, retrospectively, or prospectively. First, the longitudinal data collected from the second National Health and Nutrition Examination Survey (NHANES II) Mortality Study more convincingly imply that the relatively low plasma level of vitamin C might be closely related with a high risk of mortality from coronary heart disease (Simon et al. 2001). Simon and his colleagues also examined the data from an earlier NHANES II done from 1976 to 1980, which showed that a 0.5 mg/dL increase in plasma ascorbic acid concentration was associated with an 11% decrease in self-reported coronary heart disease (defined as angina and myocardial infarction) and stroke prevalence (Simon et al. 1998). Furthermore, the data collected in the NHANES III, a cross-sectional survey based on a probability sample of the US population conducted between 1988 and 1994 by the National Center for Health Statistics (Anonymous 1994), should be examined. The survey has examined the relation of vitamin C to a couple of health conditions, among which the relation of vitamin C to nontraditional cardiovascular disease risk factors will be reviewed. Epidemiological studies of vitamin C effects on coronary heart disease have examined the relation between those diseases and the plasma levels of vitamin C or oral intake amount of vitamin C, whose results are summarized in Tables 2.1 and 2.2, respectively.

Relation Between the Plasma Vitamin C Level and Atherosclerotic Vascular Diseases

A total of nine studies, which all reported a relation between the plasma levels of vitamin C and atherosclerotic vascular diseases, were reviewed. Among the first three studies, in which the relation of the plasma vitamin C levels to cardiovascular disease was examined, two studies showed a significant association between the plasma vitamin C level and cardiovascular disease mortality. According to Simon and Hudes's findings, 6075 participants including both males and females who were 30–75 years of age were divided into 3 groups according to their plasma levels of vitamin C (high-saturated group, $\geq$1.1 mg/dL; normal group, 0.5–1.0 mg/dL; and marginal group, <0.4 mg/dL) and then observed for 12–16 years. Interestingly, the high-saturated and normal group showed a 34% and 33% decreased risk of dying from cardiovascular disease, respectively, compared to the marginal group (Simon et al. 2001). In another study, in which almost 20,000 British participants (8860 males and 10,636 females) were followed up for 2–6 years, Khaw et al. found a more convincing association between the serum vitamin C level and cardiovascular disease mortality (Khaw et al. 2001). However, in only one study that examined more than 7000 American people (3347 males and 3724 females) aged 30–75 years, there was no significant association between the plasma vitamin C level and the mortality risk of cardiovascular disease in males, but it did find a weak association between them in female patients (Loria et al. 2000).

Table 2.1 Prospective cohort studies examining serum or plasma ascorbate concentrations and cardiovascular disease (CVD)

References	Cohort	Gender	Age range (y)	Cohort size	Follow-up (y)	Endpoints (ECD-9)	Exposure (mg/d)	RP	95%CI
CVD									
Simon et al. (2001)	USA	Both	30–75	6035	12–16	309–459	<0.4	1.0	
							0.5–1	0.67	0.5–0.9
							1.1–2.7	0.66	0.4–0.99
Khaw et al. (2001)	UK	Men	45–79	8860	2–6	400–438	0.4	1.0	
							0.7	0.9	0.6–1.5
							0.8	0.67	0.4–1.2
							1.0	0.29	0.2–0.6
							1.3	0.29	0.2–0.6
		Women		10,636			0.5	1.0	
							0.9	0.41	0.2–0.9
							1.0	0.36	0.2–0.9
							1.2	0.6	0.3–1.3
							1.5	0.41	0.2–1.0
Loria et al. (2000)	USA	Men	30–75	3347	12–16	410–414,427.5,	<0.5	1.4	0.9–2.3
						428.9429.2	0.5–0.9	1.00	0.6–1.5
						430–438,	0.9–1.3	1.30	0.9–2.0
						440–444	>1.3	1.0	
		Women		3724		(Same)	<0.7	0.9	0.6–1.5
							0.7–1.2	0.9	0.6–1.4
							1.2–1.5	0.9	0.6–1.4
							>1.5	1	

(continued)

Table 2.1 (continued)

References	Cohort	Gender	Age range (y)	Cohort size	Follow-up (y)	Endpoints (ECD-9)	Exposure (mg/d)	RP	95%CI
CHD									
Khaw et al. (2001)	UK	Men	45–79	8860	2–6	410–414	0.4	1	
							0.7	1.18	0.7–2.1
							0.8	0.92	0.5–1.7
							1.0	0.35	0.2–0.8
							1.3	0.32	0.2–0.8
		Women		10,636		(Same)	0.5	1	
							0.9	0.23	0.7–0.9
							1.0	0.39	0.2–1.2
							1.2	0.32	0.1–1.0
							1.5	0.07	0.1–0.7
Gey et al. (1993)	Basel	Men	Middle age	2974	12	410–414	<0.4	1.25	NS
Gale et al. (1995)	Britain	Both	65+	730	20	410–414	<0.21	1	
							0.21–0.49	0.9	0.6–1.3
							>0.49	0.9	0.6–1.3
Sahyoun et al. (1996)	Boston	Both	60+	725	12	Heart disease	0.9	1	
						Mortality	1.0–1.5	0.51	0.3–0.9
							1.6	0.53	0.3–1.1

Fatal or nonfatal MI									
Nyyssonen et al. (1997)	Eastern Finland	Men	42–60	1605	3–8	Fatal or Nonfatal MI	<0.20	2.80	0.8–5.3
							0.2–0.58	0.87	0.4–1.9
							0.59–0.88	0.62	0.3–1.5
							0.89–1.14	0.92	0.4–2.1
							>1.14	1.0	
Cerebrovascular diseases									
Gale et al. (1995)	Britain	Both	65+	730	20	433–438	<0.21	1	
							0.21–0.49	1.1	0.7–1.7
							>0.49	0.7	0.4–1.1

Loria CM. Vitamin C status and cardiovascular disease: A review of prospective studies. Chap. 7 in "The Antioxidant Vitamins C and E". Packer L, Traber MG, Kraemer K and Frei B (Editors). 2002. AOCS Press

ICD-9 International Classification of Diseases, 9th Revision, *RR* relative risk, *CI* confidence interval, *CHD* coronary heart disease, *NS* nonsignificant, *MI* myocardial infarction

Table 2.2 Prospective cohort studies examining vitamin C intake and cardiovascular disease (CVD)

References	Cohort	Gender	Age range (y)	Cohort size	Follow-up (y)	Endpoints (ECD-9)	Exposure (mg/d)	RP	95%CI
CVD mortality									
Enstrom (1994)	USA	Men	25–27	4479	13–16	390–459	<50 mg/day	1.00	0.9–1.1
							50+nosup	1.00	0.9–1.1
							50+sup	0.70	0.5–0.8
		Women		6869		(same)	<50 mg/day	1.00	0.9–1.1
							50+nosup	1.00	0.9–1.1
							50+sup	0.80	0.6–0.96
Unpublished[b]	USA	Men	30–75	3347	12–16	410–414,427.5,	<32	1.30	0.8–2.2
						428.9429.2,	32–73	1.60	1.0—2.6
						430–438,	74–139	1.20	0.7–2.11
						440–444	140+	1.00	
		Women		3724		(Same)	<29	0.80	0.5–1.4
							29–73	0.80	0.5–1.2
							74–134	0.90	0.6–1.4
							135+	1.00	
Fatal and nonfatal CHD									
Rimm et al. (1993)	U.S.	Men	40–75	39,910	4	Fatal CHD,	92	1.0	
						Nonfatal MI	149	1.0	0.8–1.3
						CABG,	218	1.2	0.9–1.5
						Angioplasty,	392	1.0	0.8–1.3
						Stroke	1162	0.9	0.7–1.2
Manson et al. (1993)	USA	Women	30–55	87,245	8	CHD	Q1	1.0	
							Q5	0.8	NS
	Wales	Men	45–59	2423	5	410–414,	<34.7	1.50	NS
						MI, angina	34.8–43.5	1.30	

							43.6–52.3	1.40	
							52.4–66.4	1.30	
							≤66.5	1.00	
Klipstein-Grobusch et al. (2001)	Rotterdam	Both	55–95	4802	4	121–124	<87	1.00	
						(ICD-10)	87–126	0.9	0.58–1.37
							>126	0.88	0.56–1.38
CHD mortality									
Knekt et al. (1994)	Finland	Men	30–69	2748	12–16	410–414	≤60	1.00	
						(ICD-8)	61–85	0.90	0.6–1.2
							>85	1.00	0.7–1.4
		Women		2385			≤61	1.00	
							62–91	0.50	0.3–0.99
							>91	0.50	0.2–0.98
Pandey et al. (1995)	Chicago	Men	40–55	5397	24	410–412	21–82	1.00	
							83–112	1.03	
							113–393	0.75	
							100	0.63	0.45–1.02
Gale et al. (1995)	Britain	Both	65+	730	20	410–414	<28	1.00	
							28–44.9	0.90	0.7–1.4
							≥45	0.80	0.6–1.2
Sahyoun et al. (1996)	Boston	Both	60+	725	12	Heart	90	1.00	
						Disease	91–387	0.50	0.3–0.9
						Mortality	388	0.50	0.3–1.1
Kushi et al. (1996)	Iowa	Women	55–69	34,486	8	410,412,	<112.3	1.00	
						429.2	112.4–161.3	1.08	0.69–1.69

(continued)

Table 2.2 (continued)

References	Cohort	Gender	Age range (y)	Cohort size	Follow-up (y)	Endpoints (ECD-9)	Exposure (mg/d)	RP	95%CI
							161.4–226.7	0.85	0.53–1.37
							226.8–391.2	0.99	0.61–1.59
							>391.3	1.49	0.96–2.30
Cerebrovascular disease									
Gale et al. (1995)	Britain	Both	65+	730	20	433–438	<28	1.00	
							28–44.9	0.80	0.5–1.2
							≥45	0.40	0.2–0.6
Daviglus et al. (1997)	Chicago	Men	40–55	1843	30	430–434	22–74		
						436–438	75–97	0.87	0.60–1.26
							98–122	0.87	0.60–12.6
							123–393	0.71	0.47–1.05
Yochum et al. (2000)	Iowa	Women	55–69	34,492	11	430–438	82.4	1.00	
							138.3	0.69	0.43–1.10
							190.6	0.77	0.48–1.21
							280.9	0.70	0.42–1.13
							678.7	1.24	0.81–1.89

Loria CM. Vitamin C status and cardiovascular disease: A review of prospective studies. Chap. 7 in "The Antioxidant Vitamins C and E". Packer L, Traber MG, Kraemer K and Frei B (Editors). 2002. AOCS Press

ICD-9 International Classification of Diseases, 9th Revision, *RR* relative risk, *CI* confidence interval, *sup* supplement use, *CHD* coronary heart disease, *MI* myocardial infarction, *CABG* coronary artery bypass grafting, *Q* quintile, *NS* nonsignificant

[b] Data from Loria. Unpublished

There were four studies, in which the relation between the plasma vitamin C level and the mortality risk of coronary heart disease was examined. Khaw et al. found a significant association between the blood vitamin C level and the mortality risk of coronary heart disease during a 4-year follow-up through a prospective study using a huge British population (males, 8860; females, 10,636; age, 45–79 years). In another smaller study, which examined the relation between the blood vitamin C level and heart disease mortality in a relatively small group of participants (725 men and women) who were older than 60 years living in Boston, USA, Sahyoun et al. found that the participants with the highest level of vitamin C (1.0–1.5 mg/dL) showed a decreased mortality for heart disease compared to those with a normal level (1.0 mg/dL). Unfortunately, those with more than 1.6 mg/dL did not show any additional decrease in mortality (Sahyoun et al. 1996). Another two studies did not show any significant association. However, in one of those studies, Gale et al. reported a weak association between the blood vitamin C level and coronary heart disease. They examined the relation in three groups according to the blood levels of vitamin C (lowest group, <0.21 mg/dL; middle group, 0.21–0.49 mg/dL; and highest group >0.49 mg/dL) (Gale et al. 1995). It was quite natural that a weak association was observed because the blood levels of vitamin C of all the participants in this survey were below the average plasma level of vitamin C in humans ranging from 0.55 to 0.85 mg/dL. Additionally, Gey et al. found no relation between the blood vitamin C level and coronary heart disease in 2974 middle-aged male participants. However, their average blood level of vitamin C was less than 0.4 mg/dL; thus, it was not surprising that there was no correlation between them in this epidemiological survey (Gey et al. 1993). However, unfortunately, Nyyssonen et al. found no significant correlation between the blood vitamin C level and fatal or nonfatal myocardial infarction, which is the typical atherosclerotic vascular disease of the heart in a survey of 1605 men aged 42–60 years who were followed up for 3–8 years. The participants were divided into five groups according to the plasma levels of vitamin C (lowest group, <0.20 mg/dL; lower group, 0.2–0.58 mg/dL; normal group, 0.59–0.88 mg/dL; high group, 0.89–1.14 mg/dL; and highest group, >1.14 mg/dL). Compared to the lowest group, the lower and normal groups showed significant associations which, ironically, did not show any statistical consistency for the higher and highest groups. Namely, the higher group showed a weak association, and even worse, the highest group did not show any significant association (Nyyssonen et al. 1997).

There was only one study examining the relation between the blood vitamin C level and stroke, a representative one on atherosclerotic vascular diseases occurring in the central nervous system. Gale et al. found a significant association between them in a prospective study of 730 elderly British male and female participants followed up for 20 years. Although the participants generally had maintained a relatively low plasma vitamin C level, a dose-dependent association was revealed. Specifically, compared to the lowest group (<0.21 mg/dL) and lower group (0.21–0.49 mg/dL), the high group (>0.49 mg/dL) showed a significant association (Gale et al. 1995) (Table. 2.1).

Relation Between Oral Intake of Vitamin C and Atherosclerotic Diseases

There have been several epidemiological studies, especially prospective cohort studies, that have examined the relation between oral intake of vitamin C and atherosclerotic vascular diseases (Enstrom 1994; Rimm et al. 1993; Mans et al. 1993; Fehily et al. 1993; Klipstein-Grobusch et al. 2001; Knekt et al. 1994; Pandey et al. 1995; Kush et al. 1996; Daviglus et al. 1997; Yochum et al. 2000) (Table 2.2). However, those studies are quite different from the studies examining the relation between the blood vitamin C level and atherosclerotic diseases because only a few studies have reported a significant association between oral intake of vitamin C and atherosclerotic vascular diseases, while most of the studies examining the relation between the blood vitamin C level and atherosclerotic diseases showed a significant association between them.

Why is there such a big discrepancy between these groups? First, it should be noted that the measurement of the blood vitamin C level might be a more important and objective index to evaluate the true effects of vitamin C than the vitamin C oral intake amount. According to our experiment, the maximal concentration of plasma vitamin C achievable through oral intake of any amount of vitamin C is 3.5–4.4 mg/dL (200–250 μM). Moreover, 3 h after oral intake, the vitamin C level in the blood reaches its peak, and then, another 3 h later, the vitamin C level in the blood comes back almost to the level just before the oral intake indicating that the participants should take vitamin C every 6 h to observe the authentic effects of vitamin C. There were no reports in which the participants followed the basic instructions for orally taking vitamin C. In addition, all the participants joining these prospective studies were consuming less than 500 mg of vitamin C daily (more or less 100 mg daily). Too small an amount of vitamin C was taken; thus, it was quite natural for there to be very few significant associations between the oral intake of vitamin C and atherosclerotic vascular diseases.

Role of Vitamin C in the Protection of Blood Vessels from Damage by Exercise for Health Promotion

It is impossible to keep health without exercising, which can potentiate heart and lung functions and furthermore activate energy metabolism, thereby preventing metabolic disorders such as diabetes mellitus, hyperlipidemia, and hypertension. In addition, exercise can strengthen muscle power and intensify the bone density so that it usually makes people healthy and strong. Less exercise weakens the muscles resulting in even less exercise due to a lack of power for exercise which leads to a vicious cycle. There are too many advantages or benefits to count from maintaining good health through exercise. Despite them, however, exercise has a critical problem in that it inevitably generates more oxygen free radicals proportionally to the amount of oxygen consumed during exercise, especially aerobic ones. Fortunately, it is known that increased oxygen free radicals by exercise can induce an increased production of superoxide dismutase (SOD) which can minimize the damage of oxygen free radicals to some extent. Notwithstanding, the remaining ROS not quenched by SOD in athletes usually is more than those in nonathletes, who do not usually do regular vigorous exercise; thus, athletes cannot help avoiding chronic damages from ROS. This can explain the reason why, in general, athletes, who have done more and stronger physical exercises in their life times than that of nonathletes, have a relatively short life span.

Recently, the number of patients with atherosclerotic vascular disease has dramatically increased. In addition, the onset age for the disease is becoming younger. Patients 40–50 years of age are very common, and even patients in their mid-30s are also being admitted to the hospital for atherosclerotic vascular diseases. Interestingly, it is noteworthy that among them, very sincere healthy-looking people, who firmly adhere to health promotion rules and never forget to exercise, especially aerobic ones every morning, are also included. They neither smoke cigarettes nor are obese. For example, I have a good friend, with whom I have shared a friendship for a long time, who has enjoyed triathlons for a long time, which are composed of swimming 1.5 km, then bicycling 40 km, and finally running 10 km all within 3.5 h. He used to practice swimming, bicycling, and running every morning for more than 20 years. One month ago, he visited me and told me some symptoms he was experiencing such as chest tightness and radiating pain from the left axillary region to the left chin area as well as to the forearm on the medial side of the left arm and forearm during even mild exercise suggesting a myocardial infarction. Immediately, I sent him to an outpatient clinic specializing in cardiovascular disorders for a detailed physical examination. And he was hospitalized without delay and had a coronary artery angiography revealing several intimal narrows of the coronary arteries, which could not be widened by stent due to multiple narrows but was corrected by a coronary artery bypass graft (CABG) performed by a cardiac surgeon. How can these paradoxical phenomena be explained? A review of the pathogenesis of atherosclerosis can show us that some mechanism is involved in such an unexpected surprising phenomenon. Two important changes are involved in the formation of atheroma plaques. One is endothelial damage of the arteries. And the other is peroxidation of cholesterol. Once both conditions are met, the formation of an atheroma just begins. Oxygen free radicals have a critical role in meeting both conditions. It is quite natural that the increased oxygen free radicals by vigorous and persistent exercise can accelerate atheroma formation. Especially in the early morning, the peripheral blood usually has a basal level of antioxidants including vitamin C. The pharmacokinetics of vitamin C show a peak level at 3 h which comes back to basal level after another 3 h after intake with a meal. Aerobic exercises including swimming, jogging, and power walking done persistently in the early morning for a long time promote increased production of oxygen free radicals without them being quenched by any kind of antioxidants, chronically leading to more damage on the arterial endothelium. However, these damages are left behind because the peripheral blood usually has quite a small amount of various antioxidants in the early morning. Namely, there are no ways to quench the increased ROS generated during early morning aerobic exercise. Therefore, early morning aerobic exercise should never be recommended for maintaining one's health, although doing aerobic exercise is indispensable for maintaining good health. Rather, it is strongly recommended to do aerobic exercise at 2–3 h after a meal when food in the stomach is emptied into the small intestine. Thus, it is very safe to practice even vigorous exercise without risk of heart attack. To keep a sufficient level of antioxidants, especially vitamin C in the serum during exercise, the importance of taking a sufficient amount of vitamin C with every meal cannot be overemphasized. Meanwhile, it should never be forgotten that walking exercise (not aerobic exercise) cannot only be done anytime, even in the early morning, but is also strongly recommended for maintaining good health.

References

Abbey M, Nestel PJ, Baghurst PA. Antioxidant vitamins and low density-lipoprotein oxidation. Am J Clin Nutr. 1993;58:525–32.

Abbey M, Nestel PJ, Baghurst PA, Farrel PM, Roberts RJ. Vitamin E. In: Shils ME, Olson JA, Shike M, editors. Modern nutrition in health and disease. Philadelphia: Lea and Febiger; 1994.

Anonymous. Plan and operation of the third National Health and nutrition examination survey (NHANES III), (1988–1994), the National Center for Health Statistics. Vital Health Stat. 1994;I:32.

Bae S, Cho CH, Kim H, Kim Y, Kim HR, Hwang YI, Yoon JH, Kang JS, Lee WJ. In vivo consequence of vitamin C insufficiency in liver injury: vitamin C ameliorates T-cell-mediated acute liver injury in Gulo(−/−) mice. Antoxidant Redox Signal. 2013;19(17):2040–53.

Bates CJ, Walmsley CM, Prentice A, Finch S. Does vitamin C reduce blood pressure? Results of a large study of people aged 65 or older. J Hypertens. 1998;16:925–32.

Bayata E, Lau BHS. Pycnogenol inhibits generation of inflammatory mediators in macrophages. Nutr Res. 2000;20:249–59.

Beharka AA, Han SN, Adolfsson O, Wu D, Smith D, Lipman R, Cao G, Meydani M, Meydani SN. Long-term dietary antioxidant supplementation reduces production of selected inflammatory mediators by murine macrophages. Nutr Res. 2000;20:281–96.

Berger L, Gerson CD, Yu T-F. The effect of ascorbic acid on uric acid excretion with a commentary on the renal handling of ascorbic acid. Am J Med. 1977;62:71–6.

Bowie AG, O'Neil LAJ. Vitamin C inhibits NF-κB activation by TNF via the activation of p38 mitogen-activated protein kinase. J Immunol. 2000;165:7180–8.

Carr AC, Frei B. Toward a new recommended dietary allowance for vitamin C based on antioxidant and health effects in humans. Am J Clin Nutr. 1999;69:1086–107.

Curtis LK. Reversing atherosclerosis? New Eng J Med. 2009;360(11):1144–6.

Daviglus ML, Orecia AJ, Dyer AR, Liu K, Morris DK, Persky V, Chavez N, Goldberg J, Drum M, Shekelle RB, Stamler J. Dietary vitamin C, beta-carotene and 30-year risk of stroke: results from the Western Electric Study. Neuroepidemiology. 1997;16:69–77.

Devaraj S, Hirany SV, Burk RF, Jialal I. Divergence between LDL oxidative susceptibility and urinary F(2)-isoprostanes as measures of oxidative stress in type 2 diabetes. Clin Chem. 2001;47:1974–9.

Duffy SJ, Gokce N, Holbrook M, Huang N, Frei B, Keaney JF Jr, Vita JA. Treatment of hypertension with ascorbic acid. Lancet. 1999;354:2048–9.

Enstrom JE. Vitamin C intake and mortality among a sample of United States population: new results. Stuttgart: Hippocrates Verlar; 1994.

Ersoy A, Dilek K, Yavuz M, Gululu M, Yurtkuran M. The effect of vitamin C on laboratory tests in hemodialysis patients: is there a relationship between the administered vitamin C dose and serum uric acid levels? (Lett) Nephrol Dial Transplant. 1999;14:2529–30.

Fain O, Paries J, Jacquart B, Le Moel G, Kettaneh A, Stirnemann J, Heron C, Sitbon M, Taleb C, Letellier E, Betari B, Gattegno L, Thomas M. Hypovitaminosis C in hospitalized patients. Eur J Intern Med. 2003;14:419–25.

Fehily AM, Yamell JW, Sweetnam PM, Elwood PC. Diet and incident ischemic heart disease: the Caerphilly study. Br J Nutr. 1993;69:303–14.

Frei B. Ascorbic acid protects lipids in human plasma and low density lipoprotein against oxidative damage. Am J Clin Nutr. 1991;54:1113S–8S.

Gale KF, Martyn C, Winter PD, Cooper C. Vitamin D and risk of death from stroke and coronary heart disease in cohort of elderly people. Br Med J. 1995;310:1563–6.

Gey GF. On the antioxidant hypothesiswith regard to arteriosclerosis. Bibl Nutr Dieta. 1986;37:53–91.

Gey GF. Optimum plasma levels of antioxidant micronutrients: ten years of antioxidant hypothesis on arteriosclerosis. Bibl Nutr Dieta. 1994;51:84–99.

Gey GF. Vitamins E plus C and interacting co-nutrients required for optimal health. Biofactors. 1998;7:113–74.

Gey KF, Stahelin HB, Eichholzer M. Poor plasma status of carotene and vitamin C is associated with higher mortality from ischemic heart disease and stroke: base prospective study. Clin Invest. 1993;71:3–6.

Hallfrisch J, Singh VN, Muller DC, Baldwin H, Bannon ME, Andres R. High plasma vitamin C associated with high plasma HDL- and HDL2 cholesterol. Am J Clin Nutr. 1994;60:100–5.

Ignarro LJ, editor. Nitric oxide: biology and pathobiology. San Diego: Academic; 2000.

Jacob RA. Vitamin C. In: Shils ME, Olson JA, Shike M, editors. Modern nutrition in health and disease. Philadelphia: Lea and Febriger; 1994.

Jacques PF, Sulsky SI, Perrone GA, Scaeffer EJ. Ascorbic acid and plasma lipids. Epidemiology. 1994;5:19–26.

Jeong YJ, Hong SW, Kim JH, Jin DH, Kang JS, Lee WJ, Hwang YI. Vitamin C-treated murine bone marrow-derived dendritic cells preferentially drive naïve T cells into Th1 cells by increased IL-12 secretions. Cell Immunol. 2011;266(2):192–9.

Kaller A, Hartmann D, Hornig D. Steady-state turnover and body pool of ascorbic acid in man. Am J Clin Nutr. 1979;32:530–9.

Khaw KT, Bingham S, Welch A, Luben R, Wareham N, Oakes S, Day N. Relation between plasma ascorbic acid and mortality in men and women in EPIC-Norfolk prospective study: a prospective population study. European prospective investigation into cancer and nutrition. Lancet. 2001;357:657–63.

Kim H, Bae S, Yu Y, Kim Y, Kim HR, Hwang YI, Kang JS, Lee WJ. The analysis of vitamin C concentration in organs of gulo(−/−) mice upon vitamin C withdrawal. Immune Netw. 2012;12(1):18–26.

Kim H, Bae S, Kim Y, Cho CH, Kim SJ, Kim YJ, Lee SP, Kim HR, Hwang YI, Kang JS, Lee WJ. Vitamin C prevents stress-induced damage on the heart caused by the death of cardiomyocytes through down-regulation of the excessive production of catecholamine, TNF-α, and ROS production in Gulo (−/−) Vit C-insufficient mice. Free Radic Biol Med. 2013;65:573–83.

Kim JH, Jeong YJ, Hong JM, Kim HR, Kang JS, Lee WJ, Hwang YI. Chronic vitamin C insufficiency aggravated thioacetamide-induced liver fibrosis in gulo-knockout mice. Free Radic Biol Med. 2014;67:81–90.

Kim H, Kim Y, Bae S, Lim SH, Jang M, Choi J, Jeon J, Hwang YI, Kang JS, Lee WJ, Vitamin C. Deficiency causes severe defects in the development of the neonatal cerebellum and in the motor behaviors of Gulo−/− mice. Antioxid Redox Signal. 2015;23:1270–83.

Klipstein-Grobusch K, den Breeijen JH, Grobbee DE, Boeing H, Hofman A, Witteman JC. Dietary antioxidants and peripheral arterial disease: the Rotterdam study. Am J Epidemiol. 2001;154:145–9.

Knekt P, Reunanen A, Jarvinen R, Seppanen R, Heliovaara M, Aromaa A. Antioxidant vitamin intake and coronary mortality in a longitudinal population study. Am J Epidemiol. 1994;139:1180–9.

Kush LH, Folsom AR, Prineas RJ, Mink PJ, Wu Y, Bostick RM. Dietary antioxidant vitamins and death from coronary heart disease in postmenopausal women. N Engl J Med. 1996;334:1156–62.

Kushi LH, Folsom AR, Prineas RJ, Mink PJ, Wu Y, Bostick RM. Dietary antioxidant vitamins and death from coronary heart disease in postmenopausal women. N Engl J Med. 1996;334:1156–62.

Lefer AM. Prostacyclin, high density lipoproteins and myocardial ischemia. Circulation. 1990;81:2013–5.

Lind J. A treatise of the scurvy. 2nd ed. London: Millar; 1757.

Loria CM, Klag MJ, Caulfield LE, Whelton PK. Vitamin C status and mortality in US adults. Am J Clin Nutr. 2000;72:139–45.

Lubrano V, Balzan S. LOX-1 and ROS, inseparable factors in the process of endothelial damage. Free Radic Res. 2014;48(8):841–8.

Luc G, Fruchart JC. Oxidation of lipoproteins and atherosclerosis. Am J Clin Nutr. 1991;53:306S–209S.

Machlin LJ, Bendich A. Free radical tissue damage: protective role of antioxidants nutrients. FASEB J. 1987;1:441–5.

Mans JE, Gaziano JM, Hennekens CH. Antioxidants and cardiovascular disease: a review. J Am Coll Nutr. 1993;12:426–32.

Manson JE, Gaziano JM, Jonas MA, Hennekens DH. Antioxidants and cardiovascular disease. A Review. J Am Coll Nutr. 1993;12:426–32.

Nagappan A, Park HS, Park KI, Hong GE, Yumnam S, Lee HJ, Kim MK, Kim EH, Lee WS, Lee WJ, Cho MJ, Lee WK, Won CK, Cho JH, Kim GS. Helicobacter pylori infection combined with DENA revealed altered expression of p53 and 14-3-3 isoforms in Gulo−/− mice. Chem Biol Interact. 2013a;206(2):143–52.

Nagappan A, Park HS, Park KI, Kim JA, Hong GE, Yumnam S, Eun-Kim H, Lee WS, Lee WJ, Cho MJ, Lee WK, Won CK, Kim GS. Differentially expressed proteins in the liver of Gulo−/− mice following treatments with helicobacter pylori and diethylnitrosamine. J Biomed Res. 2013b;14(2):99–104.

Nakagiri A, Sunamoto M, Murakmi M. NADPH oxidase is involved in ischaemia/reperfusion-induced damage in rat gastric mucosa via ROS production: role of NADPH oxidase in rat stomachs. Inflammopharmacology. 2007;15(6):278–81.

Nyyssonen K, Porkkala-Sarataho E, Kaikkoken J, Salonen JT. Ascorbate and urate are the strongest determinants of plasma antioxidative capacity and serum lipid resistance to oxidation in Finnish men. Atherosclerosis. 1997;130:223–33.

Paduch R, Walter-Crneck A, Zdzisinska B, Szuster-Ciesielska A, Kandefer-Szerszen M. Role of ROS, MMP-2 and IL-6 in direct interactions between tumor cell spheroids and endothelial cell monolayer. Cell Biol Int. 2005;29(7):497–505.

Pandey DK, Shekelle R, Selwyn BJ, Tangney C, Stamler J. Dietary vitamin C and beta-carotene and risk of death in middle-aged men. The Western Electric Study. Am J Epidemiol. 1995;142:1269–78.

Park YM, Febbraio M, Silverstein RL. CD36 modulates migration of mouse and human macrophages in response to oxidized LDL and may contribute to macrophage trapping in the arterial intima. J Clin Invest. 2009;119:136–45.

Pasciu V, Posadino AM, Cossu A, Sanna B, Tadolini B, Gaspa L, Marchisio A, Dessole S, Cabobianco G, Pintus G, Nephrol Dial. Akt downregulation by flavin oxidase-induced ROS gereation mediates dose-dependent endothelial damage by natural antioxidants. Toxicol Sci. 2010;114(1):101–12.

Pelletier O. Smoking and vitamin C levels in humans. Am J Clin Nutr. 1968;21:1259–67.

Peters KD, Kochanek KD, Murphy SL. Report of final mortality statistics, 1996. Hyattsville: National Center for Health Statistics; 1998.

Rimm EB, Stampfer MJ, Ascherio A, Giovannucci E, Colditz GA, Willett WC. Vitamin E consumption and the risk of coronary heart disease in men. N Engl J Med. 1993;328:1450–6.

Rovira-Llopis S, Rocha M, de Pablo C, Alvarez A, Jover A, Victor VM. Is myeloperoxidase a key component in the ROS-induced vascular damage related to nephropathy in type 2 diabetes? Antioxid Redox Signal. 2013;19(13):1453–8.

Sahyoun NR, Jacques PF, Russell RM. Carotenoids, vitamin C and E, and mortality in an elderly population. Am J Epidemiol. 1996;144:501–11.

Samson MJ, Gopaul N, Davies IR, Hughes DA, Carrier MJ. Plasma F2 isoprostanes: direct evidence of increased free radical damage during acute hyperglycemia in type 2 diabetes. Diabetes Care. 2002;25:537–41.

Sargeant LA, Wareham NJ, Bingham S, Day NE, Luben RN, Oakes S, Welch A, Khaw KT. Vitamin C and hyperglycemia in the European prospective investigation into cancer-Norfolk (EPIC-Norfolk) study: a population-based study. Diabetes Care. 2000;23:726–32.

Schepers E, Glorieux G, Dhondt A, Leybaert L, Vanholder R. Role of symmetric dimethylarginine in vascular damage by increasing ROS via store-operated calcium influx in monocytes. Transplantation. 2009;24(5):1429–35.

Simon JA. Vitamin C and cardiovascular disease: a review. J Am Coll Nutr. 1992;11:107–25.

Simon JA. Chapter 6 in 'Antioxidant vitamins C and E'. In: Packer L, Traber MG, Kraemer K, Frei B, editors. Serum ascorbic acid and disease prevalence in U.S. adults: the third National Health and Nutrition Examination Survey (NHANES III). 2002.

Simon JA, Beetens JR, Herman AG. Vitamin C increases the formation of prostacyclin by aortic rings from various species and neutralizes the inhibitory effect of 15-hydroxy-arachidonic acid. Br J Pharmacol. 1983;80:249–54.

Simon JA, Schreiber GB, Crawford PB, Frederic MM, Sabry ZI. Dietary vitamin C and serum lipids in black and white girls. Epidemiology. 1993;4:537–42.

Simon JA, Hudes ES, Browner WS. Serum ascorbic acid and cardiovascular disease prevalence in US adults. Epidemiology. 1998;9:316–21.

Simon JA, Hudes ES, Tice JA. Relation of serum ascorbic acid to mortality among US adults. J Am Coll Nutr. 2001;20:255–63.

Srivastava KC. Ascorbic acid enhances the formation of prostaglandin E1 in washed human platelets and prostacyclin in rat aortic rings. Prostaglandins Leukot Med. 1985;18:227–33.

Stein HB, Hasan A, Fox IH. Ascorbic acid-induced uricosuria: a consequence of megavitamin therapy. Ann Intern Med. 1976;84:385–8.

Timini FK, Ting HH, Haaley EA, Roddy MA, Ganz P, Creager MA. Vitamin C improves endothelium-dependent vasodilation in patients with insulin-dependent diabetes mellitus. Am J Coll Cardiol. 1998;31:552–7.

Ting HH, Timini FK, Haley EA, Roddy MA, Ganz P, Creager MA. Vitamin C improves endothelium-dependent vasodilation in forearm resistance vessels of humans with hypercholesterolemia. Circulation. 1997;95:2617–22.

Toivanen JL. Effects of selenium, vitamin E and vitamin C on human prostacyclin and thromboxane synthesis in vitro. Prostaglandins Leukot Med. 1987;26:265–80.

Will JC, Byers T. Does diabetes mellitus increase the requirement for vitamin C? Nutr Rev. 1996;54:193–202.

Yochum LA, Folsom AR, Kush LH. Intake of antioxidant vitamins and risk of death from stroke in postmenopausal women. Am J Clin Nutr. 2000;72:476–83.

Yu SJ, Bae S, Kang JS, Yoon JH, Cho EJ, Lee JH, Kim YJ, Lee WJ, Kim CY, Lee HS. Hepatoprotective effect of vitamin C on lithocholic acid-induced cholestatic liver injury in Gulo(−/−) mice. Eur J Pharmacol. 2015;762:247–55.

Chapter 3
Cancer

In human history, cancer has been one of the most dreadful diseases because so many cancer patients have died despite the intensive and extensive efforts for various treatments. In January 2014, the International Agency of Research on Cancer (IARC), one of the affiliated institutes of the WHO, published the *World Cancer Report*, which was written by 250 cancer researchers from 40 countries. According to the report, although the number of the cancer patients was 12.7 million in year 2008 and 14.1 million in 2012, those numbers were expected to increase to 19.3 million in 2025, 21.6 million in 2030, and 24 million in 2035. They also anticipated cautiously that the victims of cancer would increase greatly from 8.2 million in 2012 to 13 million in 2030.

A couple of decades ago, when patients were diagnosed with cancer in a hospital, nearly all of them eventually died. At that time, the general public did not know the concept of early detection of cancer. As such, when a person was admitted to the hospital due to some symptoms suspecting cancer, cancer fully progressed to the terminal stage resulting in a low survival rate, despite the use of all the various treatment modalities, such as surgery, chemotherapy, and radiation treatment. Actually, when reviewing the history of studies on cancer treatment for the last 50 years, it can easily be seen that there has been very little progress in achieving efficacious cancer treatments. Nevertheless, nowadays, many cancer patients have been saved by the same classical modalities for cancer treatments because recently, the healthcare system for early detection or prevention of diseases has not only been developed well but also has become popular for the general public so that the possibility for early detection of cancer has become high. Because the regular checkup system for health has become popular, the number of cancer patients examined has increased rapidly. However, the actual increase is known not to be that great. In addition, survival rates have greatly improved; nevertheless, those for patients with advanced cancers still have to be improved. Therefore, the research field of new drug development for cancer treatment is still a very active field, and many trials for definitive treatments such as immunotherapy, gene therapy, and cancer stem cell therapy for even far advanced cancers have been ongoing. Considering the complexity of

W. J. Lee, *Vitamin C in Human Health and Disease*,
https://doi.org/10.1007/978-94-024-1713-5_3

carcinogenic processes, it is not believed that any single modality for cancer treatment can guarantee the complete remission of cancers because even successful drugs have been unable to avoid inducing drug resistance in the long run. Instead, the combination therapy of several treatment modalities should be recommended. Actually, many combinations among the various therapies have been under clinical uses and trials.

Causes of Cancer Formation

There have been so many causes of cancer formation. Generally, two factors are known to play key roles in carcinogenesis. One is an environmental factor, and the other is a genetic factor. The environmental factor contains a wide variety of causes such as various chemicals; infections, especially viral infections; chronic inflammation; irradiation by several types of radiation; and so on. Throughout a person's life, exposure to various chemicals is closely related with diet and respiration. Food, which essentially maintains life, contains many chemicals such as agricultural chemicals, antiseptic chemicals, several chemical condiments, and so on. Air pollution is inevitable because modern industrialization requires using more energy. The gases emitted from cars including sulfuric acid and carbon monoxide and the various toxic substances including nicotine and tar in cigarette smoke are the main causative chemical agents for cancer formation. Various kinds of reactive species such as reactive oxygen species (ROS), reactive chloride species (RCS), reactive bromine species (RBS), and reactive nitrogen species (RNS) are also included in carcinogenic chemicals (Table 3.1) (Halliwell 2006).

Infections are known to be oncogenic factors. *H. pylori* infection is a well-known example of an oncogenic bacterial infection. It is the definitive cause of a specific type of stomach cancer such as MALToma which is a low-grade B-cell lymphoma. Papilloma viral infection and hepatitis viral infection are good examples of oncogenic viral infections causing uterine cervical cancer and hepatic cancer, respectively.

Irradiation by UV rays or X-rays is also a good example of causative agents of cancer.

Biochemically, such environmental factors can induce extensive somatic mutations, which normally take place at a frequency of 1 per 10^6 base pairs of DNA in somatic cells every day but are usually repaired by normal repair systems. However, extensive somatic mutations beyond the range of the normal repair capacity cause the somatic cells to become oncogenic so that they change to altered-self cells otherwise called cancer cells. These cancer cells are kinds of non-self cells, so they tend to evoke immune responses against them, by which they can be normally removed from the body. It is very surprising to know that normally the dual systems, the repair system, and the immune system are working to protect the body from carcinogenesis. Thus, why are there cancer patients in the world? The exact mechanisms involved in carcinogenesis occurring against the dual protection systems have

Table 3.1 Kinds of reactive species

Free radicals	Non-radicals
ROS	ROS
Superoxide, $O_2^{\cdot -}$	Hydrogen peroxide, H_2O_2
Hydroxyl, $OH^{\cdot}$	Hypobromous acid, HOBr[a]
Hydroperoxyl, $OH_2^{\cdot}$ (protonated superoxide)	Hypochlorous acid, HOCl[b]
Carbonate, $CO_3^{\cdot -}$	Ozone, O_3
Peroxyl, $RO_2^{\cdot}$	Singlet $O_2{}^1\Delta g$
Alkoxyl, $RO^{\cdot}$	Organic peroxides, ROOH
Carbon dioxide radical, $CO_2^{\cdot -}$	Peroxynitrite, $ONOO^{-}$[c]
Singlet $O_2{}^1\Sigma g^+$	Peroxynitrate, O_2NOO^{-}[c]
	Peroxynitrous acid, ONOOH[c]
	Peroxmonocarbonate, $HOOCO_2$–
	Nitrosoperoxycarbonate, $ONOOCO_2$–
RCS	RCS
Atomic chlorine, $Cl^{\cdot}$	Hypochlorous acid, HOCl[b]
	Nitryl chloride, NO_2Cl[d]
	Chloramines
	Cholrine gas (Cl_2)
	Bromine chloride (BrCl)[a]
	Chlorine dioxide (ClO_2)
RBS	RBS
Atomic bromine, $Br^{\cdot}$	Hypobromous acid(HOBr)
	Bromine gas(Br_2)
	Bromine chloride (BrCl)
RNS	RNS
Nitric oxide, $NO^{\cdot}$	Nitrous acid, HNO_2
Nitrogen dioxide, $NO_2^{\cdot}$	Nitrosyl cation, NO^+
Nitrate, $NO_3^{\cdot}$	Niutroxyl anion, NO^-
	Dinitrogen tetroxide, N_2O_4
	Dinitrogen trioxide, N_2O_3
	Peroxynitrite, $ONOO^{-}$[c]
	Peroxynitrate, O_2NOO^-
	Peroxynitrous acid, ONOOH[c]
	Nitronium cation, NO_2^+
	Alkyl peroxynitrites, ROONO
	Alkyl peroxynitrates, RO_2ONO
	Nitryl chloride, NO_2Cl
	Peroxyacetyl nitrate, $CH_3C(O)OONO_2$

"ROS" is a collective term that includes both oxygen radicals and certain non-radicals that are oxidizing agents and/or are easily converted into radicals (HOCl, HOBr, O_3, $ONOO^-$, ${}^1O_2$, H_2O_2). All oxygen radicals are ROS, but not all ROS are oxygen radicals. Peroxynitrite and H_2O_2 are frequently erroneously described in the literature as free radicals, for example. "RNS" is a similar collective term that includes $NO^{\cdot}$ and $NO_2^{\cdot}$ as well as non-radicals such as HNO_2 and N_2O_4.

(continued)

Table 3.1 (continued)

"Reactive" is not always an appropriate term: H_2O_2, $NO^{\cdot}$, and $O_2^{\cdot-}$ react fast with few molecules, whereas $OH^{\cdot}$ reacts fast with almost everything. Species such as $RO_2^{\cdot}$, $NO_3^{\cdot}$, $RO^{\cdot}$, HOCl, HOBr, $CO_3^{\cdot-}$, $CO_2^{\cdot-}$, $NO_2^{\cdot}$, $ONOO^-$, NO_2^+, and O_3 have intermediate reactivities

[a]HOBr and BrCl could also be regarded as RBS

[b]HOCl and HOBr are often included as ROS, although HOCl is also a RCS

Oxidizing species formed in polluted air that are toxic to plants and animals. $NO_2^{\cdot}$ is also produced in vivo by myeloperoxidase and from $ONOO^-$ (Augusto et al. 2002). Ozone might also be produced in vivo, although the chemistry involved is unclear (Wentworth 2003)

[c]$ONOO^-$, O_2NOO^-, and ONOOH are often included as ROS but are also classifiable as RNS

[d]NO_2Cl can also be regarded as a RNS

still not been clearly elucidated. Some cancer immunologists have explained the immunological mechanism involved in carcinogenesis, which is called the "immune escape mechanism." It means that cancer cells are so clever that they can avoid immune surveillance by forming some conditions such as a change of the MHC expression or an increase of the FasL expression on cancer cells. Specifically, the processes for immune escape consist of three phases of tumor growth. The first is the elimination phase, in which several immune responses against altered-self cells, that is, tumor cells, are working, and they are removed. However, when the tumor cells are not removed completely by the normal immune surveillance, the surviving tumor cells get into the second phase, the equilibrium phase, and are functionally transformed so that they obtain the capability to survive the normal immune surveillance. This process is called "cancer immunoediting." In the last phase, the escape phase, the tumor cells make more secure conditions avoiding the immune surveillance and grow relatively free to become clinically detectable. In the long run, their host has become a cancer patient (Fig. 3.1) (Dunn et al. 2004).

The second factor of cancer formation is a genetic one, for which active studies have not been performed. In some cases, all the members of a family develop a certain type of cancer, called a cancer family, which must have a familial trend for a certain cancer. Although the exact underlying mechanism involved in familial trends has not been fully clarified, it means that the same specific genes of certain organs or tissues in the same family members are predisposed to a mutation more frequently than family members who do not show a familial predisposition.

However, many oncogenes and tumor suppressor genes, which are known to be involved in cancer formation or prevention, have been widely studied. Recently, the mutation of major oncogenes and tumor suppressor genes such as K-RAS, N-RAS, B-RAF, EGFR, BRCA1, APC, ALK, and p53 has been shown to be closely related with carcinogenesis or with the relapse of treated cancers. In addition, aberrant promotor methylation of major tumor suppressor genes such as p16, E-CAS, RASSF1A, MGMT, and so on has been reported to be evidence of tumor formation.

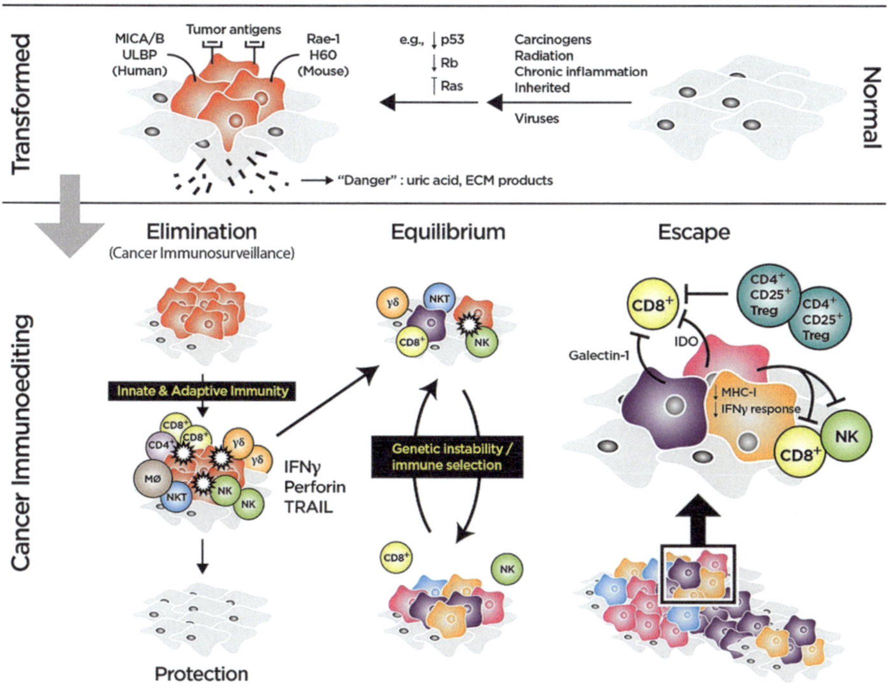

Fig. 3.1 Immune surveillance and immune escape mechanism. The immune escape process consists of three phases. (1) Elimination phase, during which the cancer cells are removed normally. (2) Equilibrium phase, during which some cancer cells are surviving the immune surveillance via immunoediting process. (3) Escape phase, after which the host has become the cancer patient

Treatment Modalities for Cancer

Various kinds of modalities for cancer treatment have been developed and used for a long time. The most classical modality is surgical treatment which is the simplest and best measure in that the tumor lesions can be completely removed in an open field although, nowadays, endoscopic surgery using a close field and even robot surgery have made surgical treatment of cancer become less invasive. Another treatment modality for a well-localized cancer is radiation therapy, which does not require direct incisions to approach the tumor lesions. Thus, it is less invasive than classical surgical treatment. Especially for cancers occurring in the head and neck area which usually prevent surgeons from treating them directly through a skin incision, radiation therapy is the treatment modality of choice. Recently, because of the development of a device with strong and specific effects without fatal damage to neighboring tissues or organs, the use of radiation therapy has increased. However, once the tumor cells start to spread to other areas, which is called metastasis, the treatment effect of surgery or radiation therapy is really limited. Therefore, once some evidence of metastasis is found, the tumor has become a kind of systemic disease, which cannot be controlled only by surgery or radiation therapy, which is the best choice of treatment for a well-localized cancer. Chemotherapy is one of the classical modalities which enables an anticancer agent to make direct contact with any tumor cell found in the human body through the extensive capillary networks. Originally, chemotherapeutic antitumor agents were developed based on the fact that tumor cells have the characteristic of uncontrolled, ceaseless proliferation, which is also normal for young developing cells. Therefore, in the early development phase of chemotherapeutic antitumor drugs, their main targets were proliferating cells, which usually had high turnover rates in the cell cycle. It is not surprising that the unwanted toxicities of the classical chemotherapeutic antitumor drugs have been usually closely related with causing damage to normal rapidly proliferating cells such as the skin (hair loss) and mucosal epithelium (nausea and vomiting) and bone marrows cells (bone marrow suppression leading to leukopenia). Nowadays, the trend for the development of antitumor chemotherapeutics is target therapy, in which the molecules involved in the oncogenic process are specifically blocked by inhibitory small molecules such as gefitinib (Iressa) for non-small cell lung cancer, imatinib (Gleevec) for chronic myeloid leukemia, or blocking monoclonal antibodies (herceptin, cetuximab, nivolumab, pembrolizumab) raised against those molecules, resulting in growth failure of the cancer cells followed by complete remission. Among the latter target therapies using blocking monoclonal antibody, the therapy by nivolumab (Opdivo) or pembrolizumab (Keytruda) is also called immunotherapy because they can block the functioning molecule (PD-1) involved in the immune suppression against the cancer cells. Generally, the therapeutic efficacies of target therapy have been greatly improved, and the unavoidable side effects of the classical chemotherapy have also been dramatically reduced. However, the use of that treatment has been strictly limited to certain types of cancers, on which the genes for the target molecules should be expressed, and even the appearance of mutant cancer cells has made its usefulness faded. Now, many other drug candidates for target therapy are under basic development and in clinical trials. Another treatment modality for systemic cancer cases is immunotherapy, which is well known to be theoretically the most ideal treatment modality among all the modalities because antibodies raised against cancer antigens can bind only to the cancer tissues specifically and kill or remove them without any damage to the surrounding normal tissue. Despite immunotherapy being theoretically perfect, there have been so many practical hurdles in the development of a successful treatment protocol. One of the representative immunotherapies is a cancer vaccine. The concept of a cancer vaccine has become the basis for a new trial to deal with the hurdles in the development of new immunotherapy technologies. There are so many immune cells responding to cancer tissues such as B cells producing antibodies that react specifically with cancer tissues, T cells killing cancer cells (cytotoxic T cells), natural killer (NK) cells also killing cancer cells directly, and dendritic cells (DC) orchestrating the functions performed by all these cells. The technology for cancer vaccines takes advantage of an inherent characteristic of vaccinations. Some components of cancer cells are used as an antigen, for which several anticancer immune mechanisms operate including cytotoxic T cells and NK cells. In the center of the anticancer mechanism, DC orchestrate all the functions of the involved cells having pivotal roles in the successful achievement of the anticancer effecter function. Now, many studies for a DC cancer vaccine are ongoing. There are even some clinical trials currently underway. In the near future, a DC cancer vaccine protocol could be developed successfully even if it is not perfect.

Another trial for the immunotherapy in the spotlight is the therapy using CAR-T (chimeric antigen receptor-T) cells, which has been known to be highly effective for patients with metastatic or relapsed cancer who have failed the traditional cancer therapy and, thus, evaluated that it was developed to meet the unmet medical demand of the cancer patients that should be solved urgently now. Nevertheless, the self-customized CAR-T-cell therapy for cancer patients has a disadvantage in that the treatment process is complicated and mass production is not possible, which is costly and uncertain. Currently, this new immunotherapy can be applied only to blood cancer such as acute lymphocytic leukemia. However, ceaseless clinical trials to treat even the solid tumors with CAR-T cells have not stopped, in spite of known adverse effects such as cytokine release syndrome and neurotoxicity.

Review of the Anticancer Effects of Vitamin C

Our unpublished data tell us that vitamin C might have some beneficial effects on cancer patient management for any of the various carcinogenic mechanisms. Vitamin C levels in the peripheral blood of liver cancer patients and healthy normal people have been compared. The average concentration of vitamin C in the 44 healthy normal people was 42.9 ± 12.4 μM, while in 83 cancer patients, it was only 17.1 ± 12.9 μM, which is at the subscurvy level (lower than 30 μM). In addition, vitamin C treatment has increased the number of lymphocytes in cancer patients from 27.0 ± 12.2% to 35.0 ± 19.2% (p-value 0.01). In stomach cancer, vitamin C treatment increased the number of NK cells dramatically, which have key roles in anticancer immunity (Dunn et al. 2004) (Fig. 3.2). Furthermore, production of some substances such as IFN-γ (Fig. 3.3) and granzyme B (Fig. 3.4), which are secreted by NK cells, increased after vitamin C treatment in stomach cancer patients.

Despite the fundamental facts suggesting some beneficial effects of cancer treatments, the anticancer effects of vitamin C have been debated for a long time. Dr. Cameron and his colleagues, a representative clinical research team investigating megadose vitamin C treatment as a cancer treatment, have reported that vitamin C, especially a megadose of vitamin C more than 10 g injected intravenously, could have many beneficial effects as an anticancer treatment such as prolongation of survival (Cameron and Campbell 1974; Cameron and Pauling 1976, 1978), improvement of a patient's sense of well-being (Cameron and Campbell 1974; Murata et al. 1982), and even complete remission of cancer. However, unfortunately, there are no clinical data based on clinical trials performed in a randomized, placebo-controlled, and double-blinded manner, which is the best way to present the results of clinical human trials. On the other hand, an oncology team at the Mayo clinic performed a comparative randomized clinical trial, in which they administered a megadose of

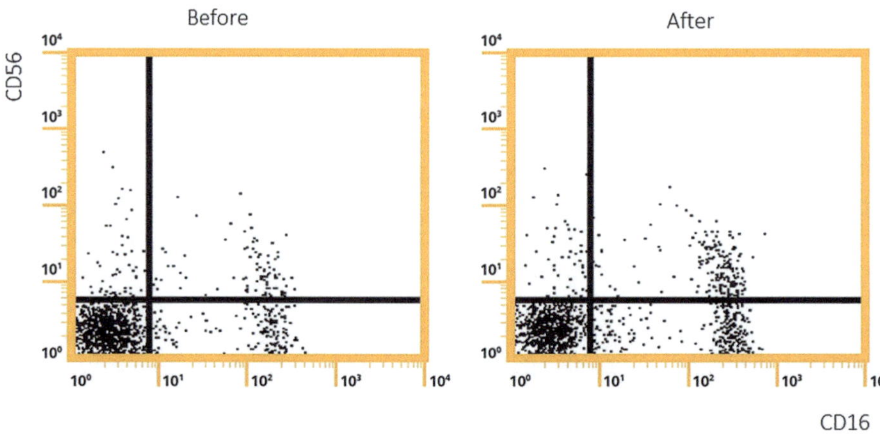

Fig. 3.2 Flow cytometric analysis for NK cells separated from mononuclear cells of stomach cancer patients before and after vitamin C treatment. The number of NK cells (CD56 and CD16 double-positive cells) increased after vitamin C treatment in stomach cancer patients (unpublished our data)

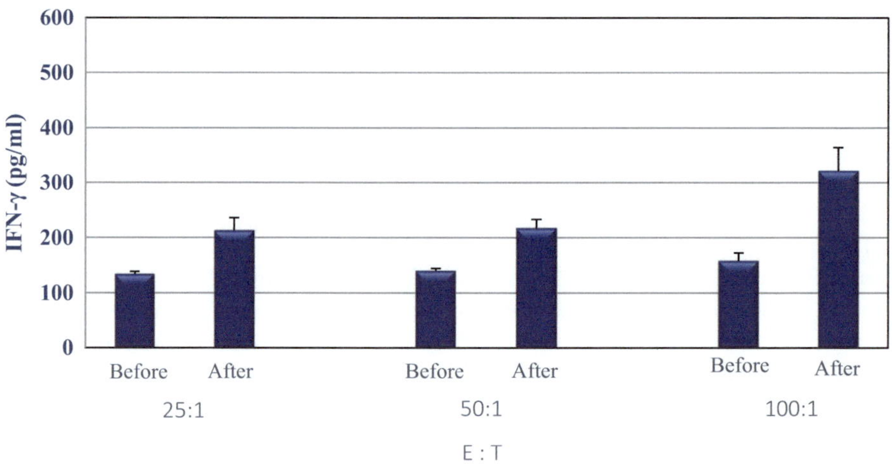

Fig. 3.3 ELISA assay for IFN-γ in the blood of stomach cancer patients before and after vitamin C treatment. Vitamin C treatment increased IFN-γ production from NK cells of stomach cancer patients. E: T; Effector (NK cells): Target (cancer cells), (unpublished our data)

vitamin C (10 g per day) to cancer patients orally; however, they did not observe any differences between the two groups, one with and one without vitamin C treatment (Creagan et al. 1979; Moertel et al. 1985). Why is there such a big discrepancy between the two teams? First, they used different routes of administration (IV injection versus oral administration) for vitamin C. Although they both used the same megadose of vitamin C, different routes of administration could produce totally different effects because maximally 200–300 μM vitamin C in the plasma was

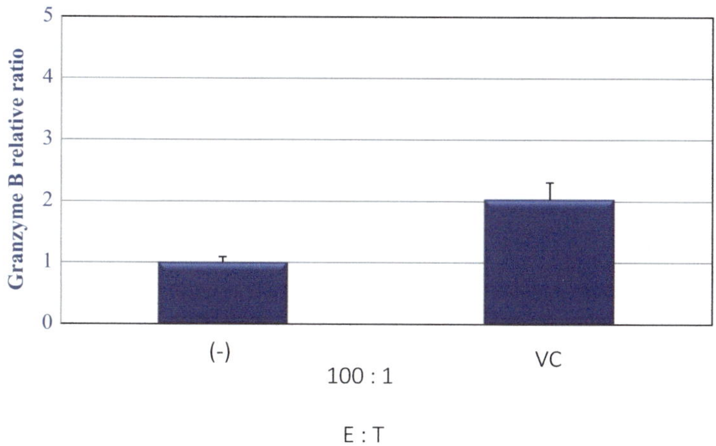

Fig. 3.4 ELISA assay for granzyme B in the blood of stomach cancer patients before and after vitamin C treatment. Production of granzyme B increased after vitamin C treatment in NK cells of stomach cancer patients. *VC* vitamin C, *N* (NK cells), *T* (Target; cancer cells) (Unpublished our data)

obtained through the oral administration, whereas vitamin C concentration in the blood reached more than 10 mM by intravenous injection, according to the amounts of vitamin C injected. My research team reported using murine melanoma cell lines and showed that a vitamin C concentration less than 500 μM could not kill the cancer cells directly. A vitamin C concentration of at least 5–10 mM could kill the cancer cells directly through apoptosis induced by the downregulation of the transferrin receptor-dependent iron uptake pathway (Kang et al. 2005). Surprisingly, it was not carried out by the antioxidant activity of vitamin C but by its prooxidant activity, which is why vitamin C is more than just your average antioxidant and is considered not to be just a typical antioxidant (Cho et al. 2003). This phenomenon was confirmed by Dr. Mark Levine and his colleagues, who showed that vitamin C induced the formation of hydrogen peroxide outside the cells, which then got into cells killing cancer cells (Chen et al. 2007). They elucidated partly why a megadose of vitamin C could not kill normal cells but could kill only cancer cells in a prooxidant manner. According to their explanation, a physiological concentration of vitamin C usually generates H_2O_2 in the extracellular fluid, which can be normally removed by plasma catalases and red blood cell GSH peroxidases (Fig. 3.5). However, cancer patients do not have plasma catalases and red blood cell GSH peroxidases; thus, they cannot quench H_2O_2 in the extracellular fluid, which then gets into the cancer cells killing them through decreased NAD+, inefficient glycolysis, and damage to mitochondria resulting in decreased ATP production. ATP may be depleted by three mechanisms: (1) DNA damage induced by H_2O_2 activates PARP. Activated PARP catabolizes NAD^+, thereby depleting the substrate for NADH formation and the consequent ATP synthesis. (2) H_2O_2 is catabolized by concurrent oxidation of GSH to GSSG. To reduce GSSG back to GSH, GSH reductase uses NADPH, which is provided by the pentose shunt from glucose. Glucose

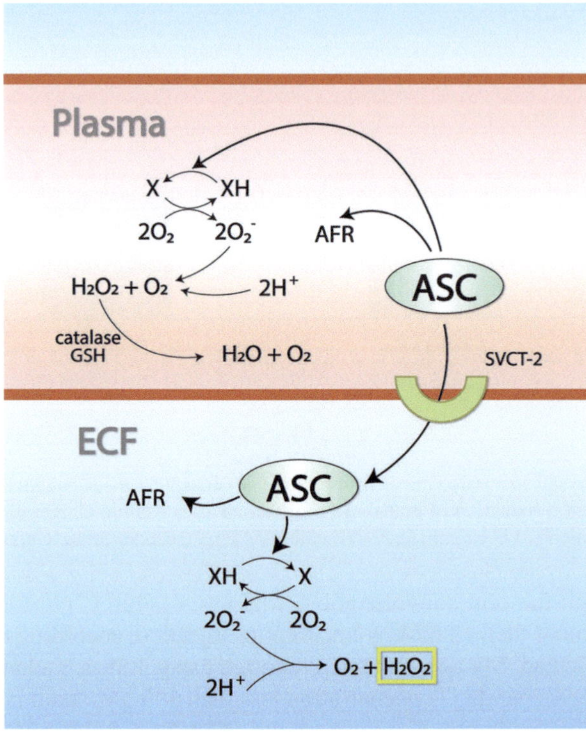

Fig. 3.5 Reason why ascorbyl free radical (AFR) and H_2O_2 are formed preferentially in the extracellular fluid compared with the blood. In the extracellular fluid, pharmacologic concentrations of ascorbic acid lose one electron and form AFR. The electron reduces a protein-centered metal: An example reaction is shown as the reduction of XH to X. X donates an electron to oxygen, forming an active oxygen including a superoxide (O2•-) with subsequent dismutation to H_2O_2, which may kill the cancer cells. However, in the blood, these reactions look like inhibited, because any formed H_2O_2 will be immediately destroyed by plasma catalase and red blood cell GSH peroxidase; thus, no H_2O_2 is detected. *ASC* L-ascorbate, *ECF* extracellular fluid. (Chen et al. 2007)

used to reduce NADP- to NADPH cannot be used for glycolysis or NADH production, and consequently, ATP generation is decreased. (3) H_2O_2 may directly damage mitochondria, especially ATP synthase, so that ATP production decreases. Some cancer cells rely primarily on glycolysis rather than on oxidative phosphorylation and respiration for ATP production (the Warburg effect). Compared with oxidative phosphorylation, ATP generation by glycolysis is inefficient. In glycolysis-dependent cancer cells, decreased glycolysis may lower the intracellular ATP. Cancer cells that are glycolysis-dependent may be particularly sensitive to pharmacologic ascorbic acid concentrations compared with cells that use oxidative phosphorylation (Chen et al. 2007). This mechanism for the vitamin C-induced cancer cell death pathway has been supported by a report by Uetaki and his colleagues. They showed that the levels of upstream metabolites in the glycolysis pathway and tricarboxylic acid (TCA) cycle were increased in MCF human breast adenocarcinoma and HT29 human colon cancer cells after exposure to a cytotoxic concentration of vitamin C, while the adenosine triphosphate (ATP) levels and adenylate energy charges were decreased in a concentration-dependent manner. Conclusively, their data suggest that vitamin C inhibited energy metabolism through NAD depletion, thereby inducing cancer cell death (Chen et al. 2007).

Furthermore, my research team has reported very important results which could be a practical guideline for the oral intake of a megadose of vitamin C for the purpose of health maintenance. In this experiment, murine melanoma cells were treated with a relatively high dose (250 μM) of vitamin C which is the maximum serum concentration of vitamin C obtained by an oral administration of a megadose of vitamin C. Surprisingly, the relatively high dose of vitamin C could not kill the cancer cells directly but did suppress their autologous production of IL-18 thereby preventing the cancer cells from immunological escaping resulting in the successful removal of the cancer cells by immune surveillance (Cho et al. 2003). These results suggest that the oral intake of a megadose of vitamin C might not have direct cancer cell killing effects but can prevent cancer formation by preventing newly formed cancer cells from immunologically escaping.

When comprehensively considering these results, it is no wonder that no anticancer effects of megadose vitamin C administered orally were observed by the oncology team at the Mayo Clinic. Furthermore, the Mayo team had cancer patients take a single dose of vitamin C. At least they should have taken a megadose of vitamin C three times per day (every 6 h) because vitamin C administered orally is to be metabolized every 6 h.

One more thing to clarify about the controversy of the anticancer effects of megadose vitamin C is whether a high dose of vitamin C could have invariably successful anticancer effects on all kinds of cancer types. The answer is no. However, it is quite evident that some cancers could be treated by a megadose of vitamin C, either completely or incompletely, while it would not be effective for other cancers at all. The scientific basis for how the differential anticancer effects of vitamin C work has not been clearly elucidated yet. Recently, my research team has reported a very important clue verifying the reason for the differential anticancer effects using various breast cancer cell lines (Hong et al., 2013). According to this report, the breast cancer cell lines strongly expressing SVCT-2 on their surface could be killed effectively by the megadose vitamin C, whereas those cell lines with low or no expression of SVCT-2 were very resistant to the treatment by the megadose vitamin C. They even provided clear evidences for this phenomenon that knocking down the SVCT-2 gene in the high-expressing SVCT-2 cell line, which made these cells resistant to the megadose vitamin C treatment, decreased chemosensitivity to vitamin C treatment, while a knock-in of the gene in a low SVCT-2 expressing cell line increased the chemosensitivity of these cells to the megadose vitamin C treatment (Hong et al. 2013) (Fig. 3.6). Furthermore, they verified these results by performing an in vivo xenograft experiment in which the size of the tumor xenograft from a high SVCT-2 expressing cancer cell line could be reduced dramatically by an intraperitoneal injection of a megadose of vitamin C in nude mice, which are known to synthesize very little vitamin C in their body, while a tumor xenograft from a low

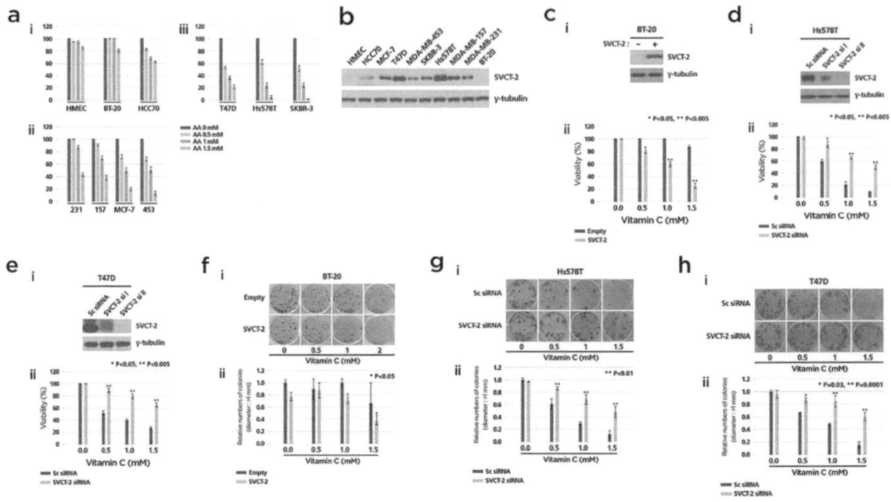

Fig. 3.6 The effects of L-ascorbate (vitamin C) according to the SVCT-2 protein level. (**a**) Effect of endogenous SVCT-2 and L-ascorbate on the cell viability of HMECs and breast cancer cell lines. Some cancer cells were chemosensitive to vitamin C treatment, while other cells were chemoresistant. (**b**) Western blot analysis of endogenous SVCT-2 protein expression in HMEC and breast cancer cell lines showed the chemosensitive cell lines were SVCT-2 high, while the chemoresistant ones were SVCT-2 low. (**c**) Transfection of SVCT-2 gene to BT20 chemoresistant cell line made them chemosensitive to vitamin C treatment. (**d**) Contrary, knockdown of SVCT-2 gene from Hs578T and (**e**) T47D cells, chemosensitive cell lines, made them chemoresistant ones. (**f–h**) Colony-forming assay performed on cells treated with or without L-ascorbate confirmed those results observed in (**c**, **d**, and **e**) were correct. (Hong et al. 2013)

SVCT-2 expressing cancer cell line showed no difference between the megadose vitamin C treatment group and the control group. These results might have clinical significance because important scientific evidence for vitamin C treatment against cancer has been presented. These results present a new possibility to save triple-negative breast cancer patients using megadose vitamin C treatment, for whom even modern anticancer medicine currently does not have any countermeasures, when considering another result reported in this experiment was that more than 60% of triple-negative patients showed positivity for the SVCT-2 antigen in the immunohistochemical staining. Although clinical trials are needed to confirm these results, now, the scientific basis for megadose vitamin C treatment of some breast cancers, especially triple-negative patients, has been settled (Hong et al. 2013) (Fig. 3.7).

Mechanisms Involved in Tumor Cell Killing by Vitamin C

According to the results of my research team through in vitro experiments using murine melanoma cells, interestingly, megadose vitamin C inhibited the proliferation of tumor cells or clearly induced apoptosis, for which there are a variety of

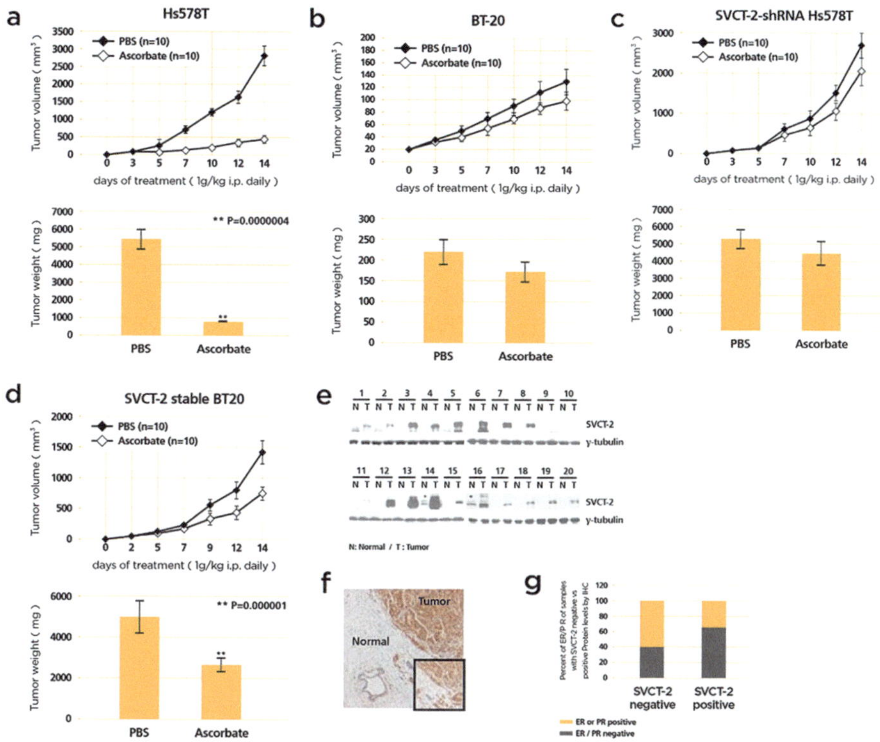

Fig. 3.7 In vivo effects of L-ascorbate in Hs578T and BT-20 cells. After (**a**) Hs578T- or (**b**) BT20-derived tumors were grown in nude mice to 2–40 mm3 in size, the nude mice were treated with L-ascorbate at a final dose of 1 g/kg by intraperitoneal injection once a day. The graph showed that vitamin C treatment to Hs578T, SVCT-2 high cell line inhibited tumor growth successfully, while the same treatment to BT-20, SVCT-2 low cell line failed to inhibit tumor growth. (**c**) Knockdown of SVCT-2 gene from Hs578T showed failure of inhibition of tumor growth, while (**d**) transfection of SVCT-2 gene to BT-20 showed successful inhibition of tumor growth. (**e**) The immunoblotting using the anti-SVCT-2 antibody for human breast normal and tumor tissues showed that most of the breast cancer tissues expressed SVCT-2. (**f**) Immunohistochemical staining against SVCT-2 in normal breast cells and tumor cells showed that SVCT-2 was highly expressed in tumor cells only. (**g**) The statistical correlation between the expression of SVCT-2 and ER/PR in human breast cancer showed that more than half among SVCT-2-positive breast cancer patients were ER/PR negative, to which the pharmacological dose of vitamin C should be a treatment of choice, because they had no measures except for the first-generation anticancer treatment modality. (Hong et al. 2013)

mechanisms: the regulation of the cell cycle, specifically G1 arrest (Hahm et al. 2007) (Fig. 3.8), the apoptotic pathway by decreasing the mitochondrial membrane potential followed by cytochrome C release in a caspase-8-independent manner (Kang et al. 2003) (Fig. 3.9), the modulation of the immune escape mechanism by decreasing the autologous production of IL-18 (Cho et al. 2003) (Fig. 3.10), and the downregulation of the transferrin receptor expression followed by a failure of iron uptake by cancer cells (Kang et al. 2005) (Fig. 3.11).

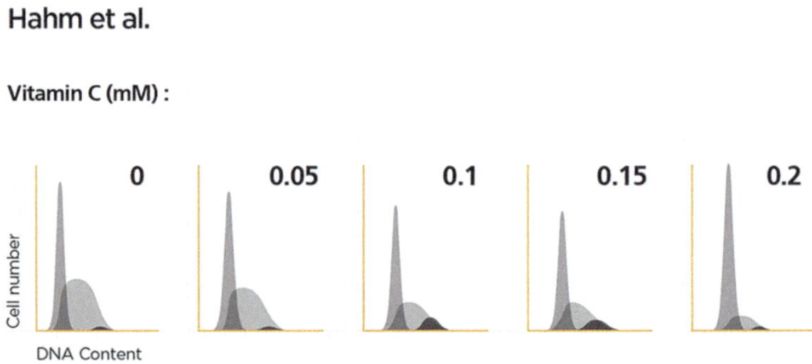

Vitamin C (mM)	Cell Cycle Distribution (%)		
	G0/G1	S	G2/M
0.00	48.0	44.1	7.9
0.05	48.0	48.7	3.4
0.10	46.0	45.8	8.2
0.15	54.9	28.5	16.6
0.20	74.6	20.0	5.4

Fig. 3.8 Flow cytometry for cell cycle in B16F10 murine melanoma cells after relatively high dose of vitamin C treatment. Melanoma cells treated with a relatively high dose of vitamin C (0.2 mM) were arrested in the G0/G1 stage of cell cycle (74.6%). (Hahm et al. 2007)

Upregulation of IL-18 production (Kang et al. 2009) by enhanced vascular endothelial growth factor (VEGF) has been known to be closely related with the metastasis of gastric cancer cells (Kim et al. 2007). Moreover, in breast cancer cells, upregulation of IL-18-induced transferrin promotes cancer cell proliferation (Park et al. 2007). Therefore, induction of the downregulation of IL-18 by megadose vitamin C can inhibit the growth and metastasis of cancer cells and even kill them. Besides apoptosis as a cell death pathway induced by vitamin C, a different type of tumor cell death, autophagy, was also induced by megadose vitamin C treatment in human breast cancer cell lines (Hong et al. 2013) (Fig. 3.12).

Nowadays, very interesting results on the antitumor effects of vitamin C have been reported. An oxidized form of vitamin C, dehydroascorbate (DHA), passively gets into cells through glucose transporter 1 (GLUT1) which is highly expressed on KRAS- and BRAF-mutant colorectal cancer cells. It induces the depletion of glutathione (GSH) and NADPH resulting in increased ROS followed by the direct and indirect (through an increase of PARP followed by the depletion of NAD+) inhibition of GAPDH, which kills KRAS- and BRAF-mutant cancer cells due to the inhibition of glycolysis and the decreased ATP. It means that megadose vitamin C treatment can cause some cancer cells to starve to death (Yun et al. 2015). This study might provide a way to deal with colorectal cancer resistant to mAb-based target therapy such as cetuximab.

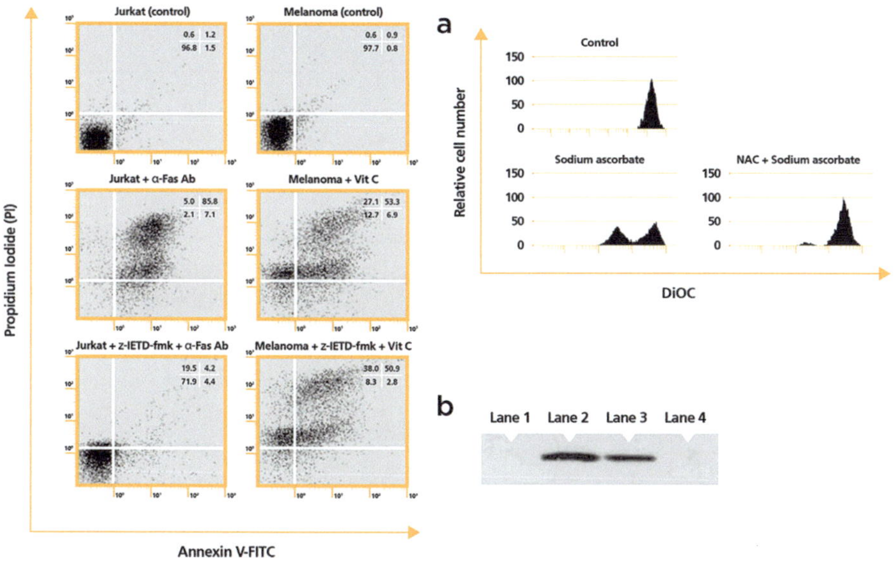

Fig. 3.9 Molecular mechanisms involved in vitamin C-induced apoptosis of B16 murine melanoma cells. Flow cytometric analysis (left) showed that vitamin C could induce apoptosis and its apoptosis was mediated in a caspase-8-independent manner. (**a**) Measurement of the mitochondrial membrane potential for B16F10 melanoma cells showed that mitochondrial membrane potential was reduced by treatment of 10 mM vitamin C and its reduction was inhibited by treatment of 10 mM N-acetyl-L-cysteine, indicating that vitamin C could induce apoptosis in a pro-oxidant, but not antioxidant manner. (**b**) Cytochrome c release assay showed that cytochrome c was released by vitamin C treatment (Lane 3) just like treatment of anti-Fas Ab (Lane 2). (Kang et al. 2003)

Despite the different mechanisms for the tumor killing processes by vitamin C, it is very important to realize that the death of all those tumor cells, regardless of the pathways used, is induced by the pro-oxidant property of vitamin C not by its antioxidant property (Uetaki et al. 2015; Cho et al. 2003; Chen et al. 2007, 2008; Hong et al. 2013; Kang et al. 2003; Gonzalez et al. 1998; Jampel 1990; Tsao et al. 1988, 1989; Bram et al. 1980). However, unfortunately, some scientists (Heany et al. 2008) without understanding the antitumor function of vitamin C as a prooxidant have reported that megadose vitamin C treatment against malignant cancers could hinder the antitumor treatment effects of antineoplastic drugs that are used clinically so far, which are known to produce mitochondrial dysfunction including the loss of the mitochondrial membrane potential and an increase in ROS levels, because they have thought that vitamin C is just one of the many antioxidants available but not a prooxidant.

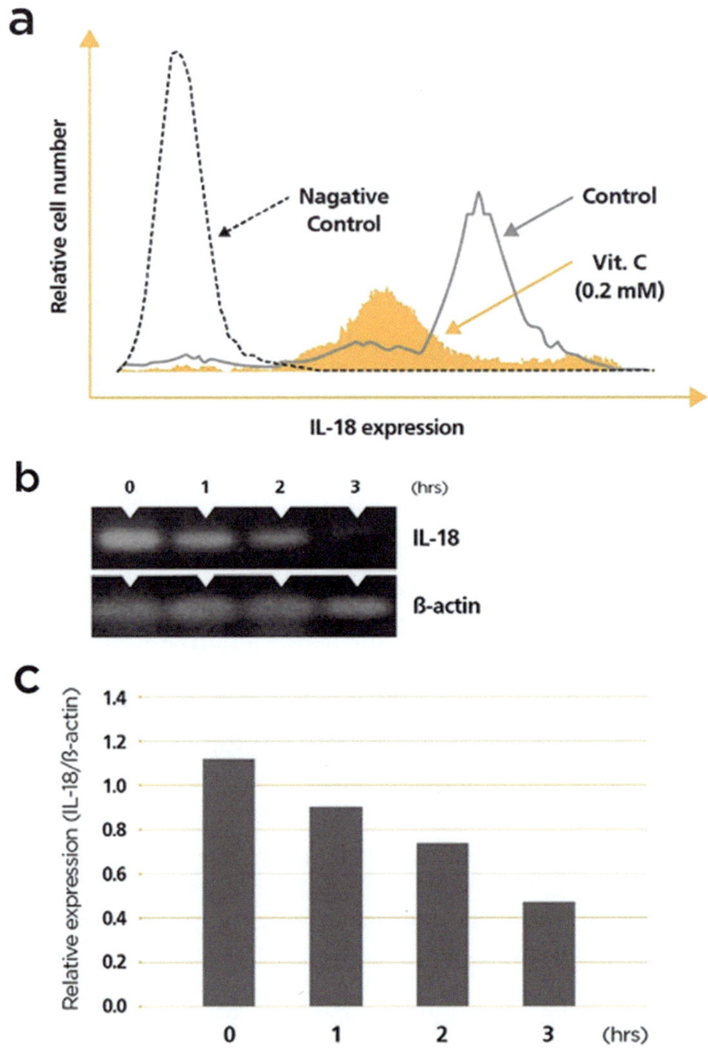

Fig. 3.10 Relatively high dose of vitamin C inhibits the production and mRNA expression of IL-18 in B16F10 melanoma cells. (**a**) Intracellular flow cytometry analysis for B16F10 melanoma cells after treatment of 0.2 mM vitamin C for 24 h showed that IL-18 production was reduced by vitamin C treatment. (**b**) RT-PCR performed after vitamin C (10 mM) treatment for 1, 2, or 3 h; mRNA expression of IL-18 was downregulated by vitamin C treatment. (**c**) The ratio of IL-18 to β-actin measured after vitamin C treatment was decreased with time after treatment. (Cho et al. 2003)

Suggestions

Among the effects of megadose vitamin C treatment in tumor patients, the improvement of the sense of well-being of patients has been reported to be the most consistent and conspicuous effect, which is good enough to be a rationale for the use of

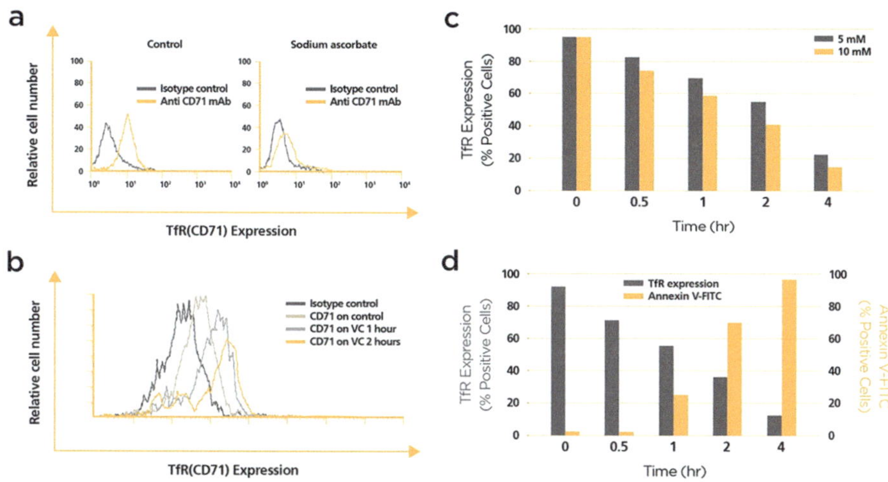

Fig. 3.11 Transferrin receptor (TfR) expression was reduced by vitamin C treatment. (**a**) FACS analysis for TfR expression after treatment of B16 murine melanoma cells with 10 mM sodium ascorbate for 4 h showed TfR expression was reduced by treatment of vitamin C. (**b**) FACS analysis for human monocytes after treatment with vitamin C for 1 and 2 h showed that TfR expression was decreased with time after treatment. (**c**) FACS analysis for human monocytes after vitamin C treatment showed that TfR expression was decreased in a time- and dose-dependent manner. (**d**) FACS analysis for human monocytes after vitamin C treatment showed that the TfR expression (black bars) was decreased in a time-dependent manner and induction of apoptosis (yellow bars) was increased in a time-dependent manner. (Kang et al. 2005)

megadose vitamin C in cancer patients even if no curative effects are exerted. Therefore, intravenous administration of megadose vitamin C to terminal cancer patients, who are in a hospice and usually suffer from intractable pain, is highly recommended because quality of life for terminal cancer patients in some cancer cases is more important than the treatment itself as a cure.

Besides the improvement in the sense of well-being of patients, there are many reasons why megadose vitamin C treatment should be recommended to cancer patients because it would have several beneficial effects when administered to them. The current antineoplastic drugs, whatever their modes of treatment are, could barely be a salvation for people in whom far advanced cancers are diagnosed. Moreover, a fact that should not be overlooked is that antineoplastic drugs usually have intolerable side effects, which make a patient's life miserable. Because megadose vitamin C treatment can also make the general condition of patients better, it can give patients the power to endure the unbearable side effects. Because megadose vitamin C treatment does not have any unwanted side effects, combination therapy with conventional chemotherapy is highly recommended. Actually, a report has been published (Ma et al. 2014) in which a high dose of parenteral ascorbate enhanced the chemosensitivity of ovarian cancer and reduced the toxicity of the chemotherapy. To deal with the weak point of anticancer-targeted chemotherapy, the formation of mutants, which are usually resistant to mAb-based target therapy, several studies on anticancer treatment including the combination therapy of vita-

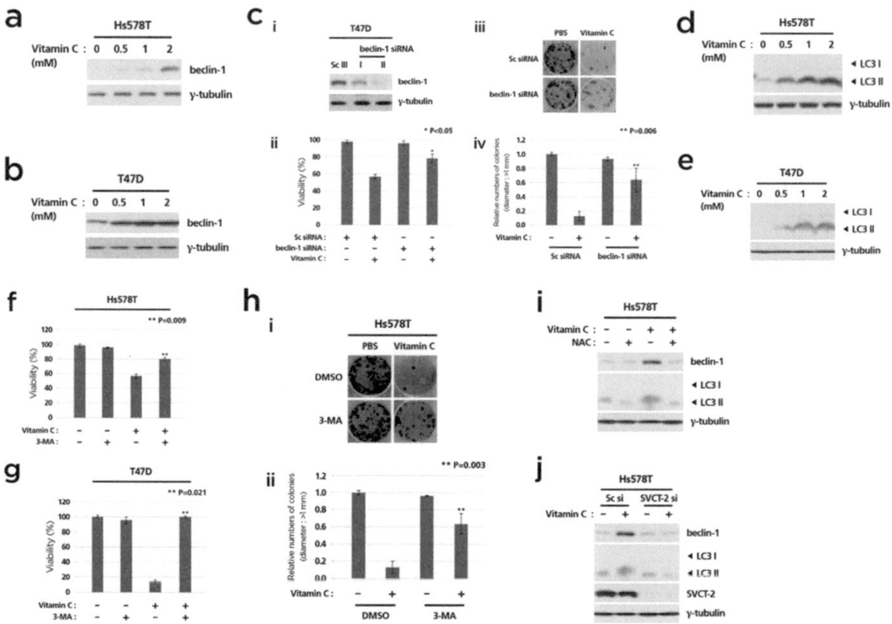

Fig. 3.12 Beclin-1 and LC3II were involved in cell death induced by L-ascorbate. The immunoblotting with the beclin-1 antibody for (**a**) Hs578T or (**b**) T47D cells showed that beclin-1 expression was increased after vitamin C treatment in a dose-dependent manner. (**c**(i)) The immunoblotting using an antibeclin-1 antibody for T47D cells after knockdown of beclin-1 gene showed that beclin-1 gene expression was decreased and (**c**(ii)) cell viability decreased by vitamin C treatment was recovered by knockdown of beclin-1 gene. (**c**(iii, iv)) Colony formation assay of the cells performed showed the same pattern of results as (**c**(ii)). (**d**, **e**) The immunoblotting with the LC3 antibody for Hs578T and T47D cells after treatment of vitamin C at the indicated doses showed that expression of LC3II was increased by vitamin C treatment. (**f**, **g**) Cell viability performed for Hs578T and T47D cells after pretreatment with 3-MA, an autophagy inhibitor, for 1 h before treatment with 0.5 mM L ascorbate showed that cell death induced by vitamin C treatment was inhibited by pretreatment of 3-MA. (**h** (i, ii)) Colony formation assay performed showed the same results as those of (**f**, **g**). (**i**) The western blot analysis for cells pretreated with NAC (10 mM) showed that expression of beclin-1 and LC3II was inhibited by treatment of NAC, indicating that vitamin C should play a role in their cell death as pro-oxidant. (**j**) Knockdown of SVCT-2 gene from Hs578T cells downregulated expression of beclin-1 and LC3II after vitamin C treatment. (Hong et al. 2013)

min C and conventional target therapy are currently in clinical trials. Recently, Yun and her colleagues reported that KRAS- and BRAF-mutant colorectal cancer cells, which usually show chemoresistance to conventional anticancer-targeted treatment, could be killed by a high dose of vitamin C, especially the oxidized form of vitamin C DHA, and showed its molecular mechanism (Yun et al. 2015) (Fig. 3.13).

Conclusively, the anticancer effects of vitamin C can be summarized in two ways: Oral administration of a relatively high dose of vitamin C could prevent cancer formation by inhibiting the immune escape mechanism of cancer cells through the suppression of IL-18 production by the cancer cells. Intravenous administration

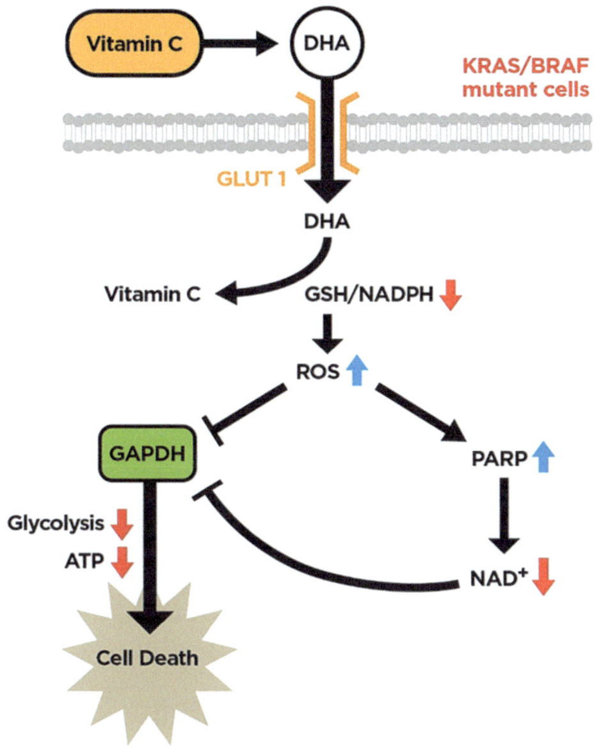

Fig. 3.13 Scheme of molecular mechanism involved in the cell death induced by high-dose vitamin C treatment of KRAS- and BRAF-mutant colorectal cancer cells. DHA is usually taken up through GLUT1, and increased DHA in cancer cells induces oxidative stress through depletion of glutathione, which occurs when intracellular DHA is reduced to vitamin C. Thereafter, ROS increases and inactivates glyceraldehyde 3-phosphate dehydrogenase (GAPDH). Inactivation of GAPDH in the highly glycolytic KRAS or BRAF mutant cells leads to an energetic crisis and cell death not seen in the KRAS and BRAF wild-type cells. Therefore, high-dose vitamin C treatment in the oxidized form of vitamin C, DHA, might be able to impair tumor growth in Kras G12D-mutant mice

of a high dose of vitamin C up to 100 g could kill cancer cells directly and be used as one modality for the curative treatment of cancer. The treatment outcome could even be better if this modality is combined with several conventional treatment modalities. This suggestion has been proven to be the most ideal way to treat cancer, which is an inevitable disease of modern people that has not been cured yet. An oncology team working in the Holden Comprehensive Cancer Center at the University of Iowa has reported striking effects from the combination therapy of conventional chemotherapy plus irradiation therapy and pharmacological ascorbate therapy on non-small cell lung cancer (NSCLC) and glioblastoma multiforme (GBM), which are clinically the most malignant cancers. Through an elaborate pre-clinical study, they have elucidated a substantial mechanism involved in the actually

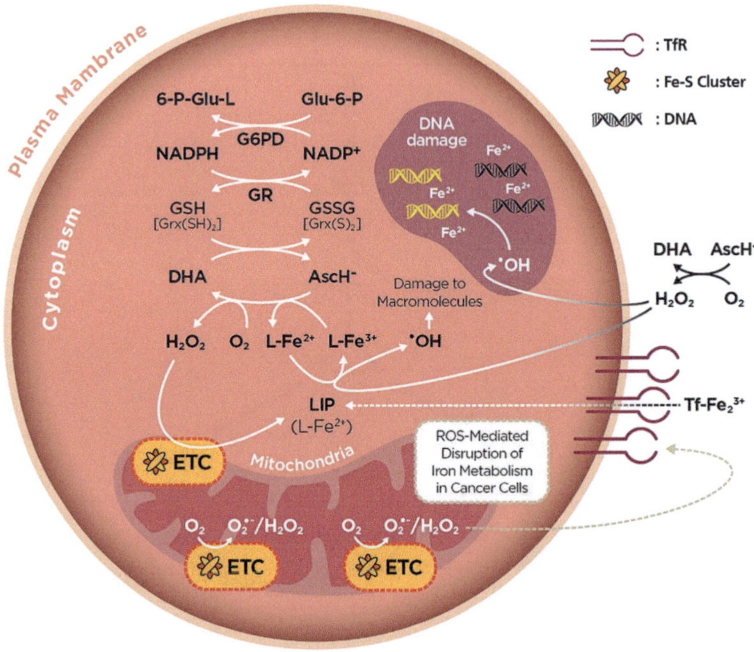

Fig. 3.14 Schematic mechanism of cancer cell-selective toxicity by high dose of vitamin C. The dashed lines represent multistep processes not illustrated in the model. Briefly, the increased steady-state levels of mitochondrial O_2 and H_2O_2, which are capable of disrupting cellular iron homeostasis, increase the levels of redox-active labile iron in the cancer cells. H_2O_2 produced by oxidation of vitamin C reacts with the increased labile iron pool in the cancer cells to mediate the Fenton chemistry and induce oxidative stress to cellular macromolecules, such as DNA, proteins, and lipids. The diffusion-limited kinetics of the HO species makes the redox-active iron chelated by these macromolecules most likely represent the most prevalent site of damage. In addition, partially by disrupting Fe-S clusters, H_2O_2 produced by vitamin C oxidation selectively increases the cancer cell labile iron pool, aggravating the differences in labile iron pool available for oxidation reactions, and mediates vitamin C toxicity in cancer versus normal cells. In this model, endogenous vitamin C recycling mechanisms are driven by reducing equivalents from nicotinamide adenine dinucleotide phosphate (NADPH) and glutathione (GSH) and/or glutaredoxin (Grx) enabling the continuous production of H_2O_2. *Glu* glucose, *G6PD* glucose6-phosphate dehydrogenase, *GR* glutathione reductase, *L-Fe* weakly chelated redox-active iron, *Tf* transferrin. (Schoenfeld et al. 2017)

improved anticancer effects obtained by the combination therapy between the first generation of conventional chemotherapy plus irradiation therapy and the intravenous pharmacological ascorbate therapy against advanced-staged NSCLC and GBM compared with those anticancer effects obtained by first-generation chemotherapy platinum-doublet chemotherapy plus irradiation against NSCLC and temozolomide plus irradiation against GBM, respectively. They revealed that pharmacological ascorbate induced changes in the cancer cell mitochondrial metabolism by increasing the production of $O_2^{\bullet-}$ and H_2O_2, which resulted in an increased chemosensitivity to ascorbate through its pro-oxidant property (Schoenfeld et al.

2017) (Fig. 3.14). The ranges of the pharmacological ascorbate doses they used in this clinically important experiment were from 15 to 125 g per injection.

Another comprehensive review of the effects of vitamin C on cancer has been published in Immune Network (Lee 2009).

References

Augusto O, Bonini MG, Amanso AM, Linares E, Santos CCX, De Menezes Sílvia L. Nitrogen dioxide and carbonate radical anion: two emerging radicals in biology. Free Radic Biol Med. 2002;32(9):841–59.

Bram S, Frossard P, Guichard M, Jasmin C, Augery Y, Sinoussi-Barre F, Wray W. Vitamin C preferential toxicity for malignant melanoma cells. Nature. 1980;284:629–31.

Cameron E, Campbell A. The orthomolecular treatment of cancer II. Clinical trial of high-dose ascorbic acid supplements in advanced human cancer. Chem Biol Interact. 1974;9:285–315.

Cameron E, Pauling L. Supplemental ascorbate in the supportive treatment of cancer: prolongation of survival times in terminal human cancer. Proc Natl Acad Sci USA. 1976;73:3685–9.

Cameron E, Pauling L. Supplemental ascorbate in the supportive treatment of cancer: reevaluation of prolongation of survival times in terminal human cancer. Proc Natl Acad Sci USA. 1978;75:4538–42.

Chen Q, Espey MG, Sun AY, Lee JH, Krishna MC, Shacter E, Choyke PL, Pooput C, Kirk KL, Buettner GR, Levine M. Ascorbate in pharmacological concentrations selectively generates ascorbate radical and hydrogen peroxide in extracellular fluid in vivo. Proc Natl Acad Sci USA. 2007;104:8749–54.

Chen Q, Espey MG, Sun AY, Pooput C, Kirk KL, Krishna MC, Khosh DB, Driko J, Levine M. Pharmacological dose of ascorbate acid as a pro-oxidant and decrease growth of aggressive tumor xenografts in mice. Proc Natl Acad Sci USA. 2008;105:11105–9.

Cho D, Hahm E, Kang JS, Kim YI, Jun HK, Yang Y, Park JH, Kim D, Kim S, Kim YS, Hur D, Park H, Pang S, Hwang YI, Lee WJ. Vitamin C down-regulates IL-18 production by increasing ROI and MAPK signaling in B16F10 murine melanoma cells. Melanoma Res. 2003;13:549–54.

Creagan ET, Moertel CG, O'Fallen JR, Schutt AJ, O'Connell MJ, Rubin J, Frytak S. Failure of high dose vitamin C (ascorbic acid) therapy to benefit patients with advanced cancer. A Control Stud N Eng J Med. 1979;301:687–90.

Dunn GP, Old LJ, Schreiber RD. The immunobiology of cancer immunosurveillance and immunoediting. Immunity. 2004;21(2):137–48.

Gonzalez MJ, Mora E, Riordan HD, Mojica P. Rethinking vitamin C and cancer: an uptake on neutritional oncology. Cancer Prev Int. 1998;3:215–24.

Hahm E, Jin DH, Kang JS, Kim YI, Hong SW, Lee SK, Kim HN, Jung DJ, Kim JE, Shin DH, Hwang YI, Kim YS, Hur DY, Yang Y, Cho D, Lee MS, Lee WJ. The molecular mechanisms of vitamin C on cell cycle regulation in B16F10 murine melanoma. J Cell Biochem. 2007;102:1002–10.

Halliwell B. Oxidative stress and neurodegeneration: where are we now? J Neurochem. 2006;97:1634–58.

Heany ML, Gardner JR, Karasavvas N, Golde DW, Scheinberg DA, Smith EA, O'Connor OA. Vitamin C antagonizes the cytotoxic effects of antineoplastic drugs. Cancer Res. 2008;68:8031–8.

Hong SW, Lee SH, Moon JH, Hwang JJ, Kim DE, Ko E, Kim HS, Cho IJ, Kang JS, Kim DJ, Kim JE, Shin JS, Jung DJ, Jeong YJ, Cho BJ, Kim TW, Lee JS, Kang JS, Hwang YI, Noh DY, Jin DH, Lee WJ. SVCT-2 in breast cancer acts as an indicator for L-ascorbate treatment. Oncogene. 2013;32(12):1508–17.

Jampel HD. Ascorbic acid is cytotoxic to dividing human Tenon's capsule fibroblasts. A possible contributing factor in glaucoma filtration surgery success. Arch Ophthalmol. 1990;108:1323–5.

Kang JS, Cho D, Kim YI, Hahm E, Yang Y, Kim D, Hur D, Park H, Bang S, Hwang YI, Lee WJ. L-ascorbic acid (vitamin C) induces the apoptosis of B16 murine melanoma cells via a caspase-8 independent pathway. Cancer Immunol Immunother. 2003;52:693.

Kang JS, Cho D, Kim YI, Hahm E, Kim YS, Jin SN, Kim HN, Kim D, Hur D, Park H, Hwang YI, Lee WJ. Sodium ascorbate (vitamin C) induces apoptosis in melanoma cells via the down-regulation of transferrin receptor dependent iron uptake. J Cell Physiol. 2005;204:192–7.

Kang JS, Bae S, Kim H, Kim YS, Kim D, Cho BJ, Yang HK, Hwang YI, Kim K, Park HS, Hwang D, Cho D, Lee WJ. Interleukin-18 increases metastasis and immune escape of stomach cancer via the down-regulation of CD70 and maintenance of CD44. Carcinogenesis. 2009;30(12):1987–96.

Kim KE, Song H, Kim TS, Yoon D, Kim CW, Bang SI, Hur DY, Park H, Cho DH. IL-18 is a critical factor for vascular endothelial growth factor-enhanced migration in human gastric cancer cell lines. Oncogene. 2007;26:1468–76.

Lee WJ. The prospects of vitamin C in cancer therapy. Immune Netw. 2009;9(5):147–52.

Ma Y, Chapman J, Levine M, Polireddy K, Drisko J, Chen Q. High-dose parenteral ascorbate enhanced chemosensitivity of ovarian cancer and reduced toxicity of chemotherapy. Sci Transl Med. 2014;5(6):222.

Moertel CG, Fleming TR, Creagan ET, Rubin J, O'Connell MJ, Ames MM. High-dose vitamin C versus placebo in the treatment of patients with advanced cancer who have had no prior chemotherapy. A randomized double-blind comparison. N Engl J Med. 1985;312:137–41.

Murata A, Morishige F, Yamaguchi H. Prolongation of survival times of terminal cancer patients by administration of large doses of ascorbate. Int J Vitam Nutr Res Suppl. 1982;23:103–13.

Park S, Cheon S, Cho D. The dual effects of IL-18 in tumor progression. Cell Mol Immunol. 2007;4:329–35.

Schoenfeld JD, Sibenaller ZA, Mapuskar KA, Wagner BA, Cramer-Morales KL, Furgan M, Sandhu S, Carlisle TL, Smith MC, Hejleh TA, Berg DJ, Zhang J, Keech J, Parekh KR, Bhatia S, Monga V, Bodeker KL, Ahmann L, Vollstedt S, Brown H, Shanahan Kauffman EP, Shall ME, Hohl RJ, Clamon GH, Greenlee JD, Howard MA, Shultz MK, Smith BJ, Riley DP, Domann FE, Cullen JJ, Buettner GR, Buatti JM, Spitz DR, Allen BG. O2•− and H2O2-mediated disruption of Fe metabolism causes the differential susceptibility of NSCLC and GBM cancer cells to pharmacological ascorbate. Cancer Cell. 2017;31:1–14.

Tsao CS, Dunhan WB, Leung PY. In vivo antineoplastic activity of ascorbic acid for human mammary tumor. In Vivo. 1988;2:147–50.

Tsao CS, Dunhan WB, Leung PY. Effect of ascorbic acid and its derivatives on the growth of human mammary tumor xenograft in mice. Can J. 1989;5:53–9.

Uetaki M, Tabata S, Nakasuka F, Soga T, Tomita M. Metabolism alterations in human cancer cells by vitamin C-induced oxidative stress. Sci Rep. 2015;5:13896.

Yun J, Mullarky E, Lu C, Bosch KN, Kavalier A, Rivera K, Roper J, Cjio IIC, Giannopoulou EG, Rago C, Muley A, Asara JM, Paik J, Elemento O, Chen Z, Pappin DJ, Dow LE, Papadopoulos N, Gross SS, Cantley LC. Vitamin C selectively kills KRAS and BRAF mutant colorectal cancer cells by targeting GAPDH. Science. 2015;350:1391–6.

Wentworth P. Evidence for ozone formation in human atherosclerotic arteries. Science. 2003;302(5647):1053–6.

Chapter 4
Immune System

All living bodies are essentially equipped with a system to defend their body called the "immune system." The immune system consists of non-specific and specific systems. Approximately 1% of the cells (~10^{14}) that comprise our body belong to the immune system (~10^{12} cells).

Non-specific Immune System

The non-specific immune system is comprised of noncellular and cellular systems. There are so many examples of non-specific, noncellular defense systems. The representative example is the skin system, which has tight and adhering junctions between skin cells so that it can prevent any harmful microorganisms from invading the body. Even water cannot penetrate the cutaneous physical barrier despite the long exposure of the skin to water.

The central body temperature, 37.5 °C, is also a very good example of a non-specific, noncellular body defense system because the central body temperature itself can have an important role in inhibiting the proliferation of most microbial causative agents of infectious diseases including bacteria and viruses. Furthermore, when our body is invaded by any infectious agents, sometimes a high fever occurs elevating the standard central body temperature to over 40 °C or higher which is regulated by the hypothalamus. The chills mean that a person feels cold even though their central body temperature is higher than 37.5 °C, which is the normal central body temperature, because the current body temperature does not reach the high central body temperature newly established as the standard central body temperature due to invading microorganisms. These important physiologic responses for the body defenses strongly suggest that people should remember not to take medicine for fever without a physician's instructions when they suffer from a high fever caused by infectious agents. Urination is another kind of non-specific noncellular body defense. Anatomically, the urinary pathway from the kidney via the urinary

W. J. Lee, *Vitamin C in Human Health and Disease*,
https://doi.org/10.1007/978-94-024-1713-5_4

Table 4.1 Non-specific defense system

Examples	Function
Noncellular system:	
Skin	Role of mechanical barrier against all kinds of foreign substances
Sebum	Low pH (3~5) with lactic and fatty acids inhibits growth of most microbes
Temperature	Inhibitory function of growth of many microorganisms by normal body temperature (37.5 °C). Fever has even stronger inhibitory function
Mucous membranes	(1) Entrapment of microbes with mucus, (2) competition of normal flora with microorganisms for nutrients and attachment sites in GI tracts, (3) removal of foreign substances attached on mucociliary blanket by one-way beating movement of cilia in respiratory tracts
Acidic pH	Killing most microbes by HCl secreted in stomach
Urination	Flushing out any microorganisms by one-way flow of urine
Saliva	Growth inhibition of microorganism by lysozymes and lactoferrin
Soluble mediators	Antiviral function of interferon and microbial lytic function of complement
***Cellular system*:**	
Phagocytosis	Breakdown of foreign substances (including micro-organisms) by engulfment of specialized cells such as blood monocytes, neutrophils, and tissue macrophages
Inflammation	Reduction of tissue damage or fight against infection by secretion of tissue fluid and recruitment of phagocytic cells

bladder to the outlet of the urethra is open to the outside, so it is always possible for infectious microorganisms to get the inside of the body through the urinary tract system. Therefore, the direction of urine flow should be one-way, from the kidney to the urethra. It is thought that the role of urination is to wash out the urinary tract to prevent colonization of microorganisms in any local part of the urinary tract. Even urinary stasis observed frequently in pregnant women due to compression of the ureters by an enlarged uterus causes frequent urinary tract infections. Table 4.1 shows many other examples of non-specific, noncellular defense systems, which are considered as structural defense tools (Table. 4.1).

Meanwhile, there is another non-specific defense system, a cellular one, which is also called the innate immune system which is mediated by immune cells, such as macrophages, granular leukocytes (neutrophils, eosinophils, and basophils), and natural killer cells. This system immediately responds to various invasions of microorganisms commonly found around human lives. First, primarily macrophages take part in these defense responses and remove them by engulfing them. Simultaneously they secrete cytokines, which can cause vasodilation and increased vascular permeability, resulting in inflammatory responses, of which four signs are redness, heat, swelling, and pain over the area of infection. Furthermore, they recruit neutrophilic leukocytes by secreting chemokines, which can also engulf microorganisms and kill them. These processes are commonly called inflammation, which usually ends within 3 days after infection, when people start to observe a pustular lesion containing pus, which stands for victory by the host in the battle against infectious agents.

Specific Immune System

Besides this non-specific but rapid defense systems, there should be a specific but relatively long-lasting immune system, which is mediated exclusively by lymphocytes, also called agranular leukocytes. The science of immunology originated from the battle between humans and infectious diseases. In 1796, when even the concept of infection by microorganisms has not been settled scientifically, for the first time, Edward Jenner performed a so-called vaccination with cowpox against smallpox. His bold experiment, in which cow serum (containing *Vaccinia virus*) from cowpox disease, a relatively mild cow disease, was inoculated into humans to successfully protect people from smallpox, an even more severe disease. Actually, the principle of vaccination is to inoculate weakened or attenuated infectious agents in humans so that some protective mechanisms are operating, and eventually, lives are saved from fatal infectious diseases when people are infected with the same microorganisms. Those protective mechanisms are the so-called phenomena of an immunological memory, a key concept of specific immunity. The memory from less severe or mild infectious diseases of some strains confers to humans as a protective immunological capacity against fatal or more severe ones from the same strains. The basic difference between specific and non-specific cellular immune systems is that the former, the specific immune system, has an immunological memory while the latter, the non-specific cellular immune system, does not. Anyway, his bold experiment has brought about the declaration by the WHO in 1979 that the viral infectious disease, smallpox, had been eradicated from the world, which is worthy of praise as the greatest victory of modern medical science. The final goal of the immune function is the elimination of foreign substances, which are called "antigens." Early on in the immunology field, only the substance antibodies, capable of eliminating foreign substances, was thought to be the effector mechanism of the immune system. Actually, an antibody can work only with exposed antigens in aqueous humor which are located in the extracellular spaces such as in the tissues or in the peripheral blood, and they cannot recognize hidden antigens, for example, antigens in the cytoplasm, which antibodies cannot reach directly. In the latter case, to remove the hidden antigen, sacrificing the normal cells containing the antigen is unavoidable. Antibody-mediated immune responses for exposed antigens are called humoral immune responses, whereas non-antibody-mediated immune responses for hidden antigens are called cell-mediated immune responses because some immune cells such as cytotoxic T lymphocytes are needed to kill antigen-containing cells. The cells with key roles in adaptive immunity are lymphocytes, which are divided into B and T lymphocytes, named by their different origin of development, *b*one marrow for B cells and *t*hymus for T cells, respectively. After antigen challenge, B lymphocytes proliferate and differentiate into antibody-producing cells (plasma cells), and their secreted antibody can remove the challenged antigen (humoral immune response), while some T lymphocytes become cytotoxic T lymphocytes, which can kill cells containing an antigen (cell-mediated immune response). There are countless numbers of antigens in the world. In adaptive immunity, each lymphocyte has

Table 4.2 Comparison of immune cells

Cells	Innate immune cells (macrophages, dendritic cells)	NK cells	Adaptive immune cells (B and T lymphocytes)
Memory	No	No	Yes
Self-discrimination	Yes	Yes (simple)	Yes (sophisticated)
Specificity	No	No	Yes (highly specific)

receptors specific for all kind of antigens existing in human life. Once an antigen binds to specific receptor, the lymphocytes are activated, resulting in the proliferation of lymphocytes with receptors having the same specificity and finally differentiating into effector lymphocytes. However, during the development of lymphocytes, those having receptors with a specificity for self-antigens are removed by deletion; this process is called central tolerance. In case of failure or defects in this process in some people, they usually suffer from autoimmune diseases such as rheumatoid arthritis and systemic lupus erythematosus, in which normal tissues are destroyed by immune cells (effector B or T lymphocytes) attacking them as a result of the B or T lymphocytes recognizing the host's normal tissue as an antigen.

In the mid-1980s, a new kind of immune cell was discovered. There was a case report that a very young malignant cancer patient, in whom the adaptive immunity consisting of B and T lymphocytes was not fully developed, showed complete remission without any anticancer medication. The scientific efforts of some immunologists revealed that a potent anticancer function was performed by a new immune cell, "natural killer (NK) cells," which can kill cancer cells without other immune cells such as cytotoxic T cells or antibody-secreting plasma cells. These cells have a morphological similarity with lymphocytes except that they have granules in the cytoplasm. Actually, the original lymphocytes do not have granules in the cytoplasm; thus, they are called agranular leukocytes, while NK cells are called granular lymphocytes due to the presence of granules in their cytoplasm. Because NK cells do not have immunological memory nor specificity, they are not included in the adaptive immunity but in the innate immunity. Although they are included in the innate immunity, their defense mechanism against cancer cells or viral infection is so complicated and strong compared to the other cells of the innate immunity such as macrophages. Actually, NK cells are intermediate cells between innate immune cells and adaptive immune cells (Table. 4.2).

Link Between the Innate and Adaptive Immune Systems

Antigens should be processed for their successful presentation to immune cells, specifically B and T lymphocytes. Only processed antigens can be presented to them, eventually evoking immune responses. Such a presentation of antigen is carried out by professional cells, the so-called antigen-presenting cells (APC), such as macrophages, B lymphocytes, and dendritic cells, of which a representative one is macrophages, derived from monocytes in the blood after coming out from the blood vessels into the tissues. APC are usually able to take up antigens (called "phagocytosis") and then process them, presenting only part of the original antigen to the immune cells. APC usually have pattern recognition receptors on their surface such as the mannose receptor, lipopolysaccharide (LPS) receptor (CD14), Toll-like receptor (TLR)-2, TLR-4, glucan receptor, and scavenger receptor, which specifically recognize the regular patterns of molecular structures found on microorganisms known as pathogen-associated molecular patterns (PAMPs), but not on the host cells. There are several kinds of PAMPS such as mannose-rich oligosaccharides, peptidoglycans, lipopolysaccharides in the bacterial cell wall, and unmethylated CpG DNA. Recognition of pathogens through the pattern recognition receptors enables the innate immune system to initially discriminate between self and nonself.

Dendritic cells have long finger-like processes, like the dendrites of neurons, which are the reason why these cells are named as "dendritic" cells. Dendritic cells have a unique function to link the innate immunity to the adaptive one through phagocytosis, which is also used for the clearance of microorganisms, which is the main role of macrophages and neutrophils in the immune system. However, dendritic cells have one more function besides the phagocytic one. Specifically, immature dendritic cells residing in the peripheral tissue take up and degrade pathogens, migrating via lymphatic vessels to regional lymph nodes where they are transformed to fully mature phagocytic dendritic cells that express a couple of receptors necessary to activate the naïve T lymphocytes, which recognize the antigen, proliferate, and differentiate to effector lymphocytes that have key roles in the adaptive immune system.

Vitamin C and B Lymphocytes

B lymphocytes have a pivot role in the immune functions as one critical axis of the adaptive or acquired immune system. They produce "antibody (Ab)" as an effector tool, which is an immunoglobulin molecule biochemically. The immunoglobulin molecule consists of a light and heavy polypeptide chain, both of which form a peptide complex by disulfide linking. Two same complexes actually comprise one Y-shaped dimerous immunoglobulin molecule through the disulfide bond. One immunoglobulin molecule has two antigen (Ag)-binding sites (upper two ends of the Y letter), called the "variable regions," responsible for the amino-terminal half of the chain, and one anchoring site (the bottom end of the Y letter) to the surface of B lymphocytes, called the "constant' region, the carboxyl-terminal half of the chain, which is responsible for the biological functions of the Ab. There are five isotypes of immunoglobulins, according to different heavy chain-constant region sequences: IgM (containing μ heavy chain), IgD (containing δ heavy chain), IgG (containing γ

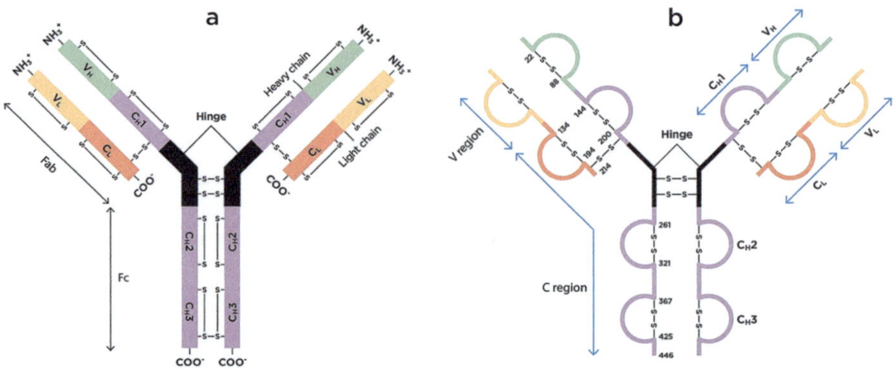

Fig. 4.1 Schematic structure of immunoglobulin molecule. (**a**) Each chain (heavy and light) has variable (V) regions at the amino terminals and constant (C) regions at the carboxyl terminals. (**b**) A heavy chain (H) consists of one variable domain and three to four constant domains, while a light chain (L) consists of one V domain and one C domain. The amino terminal domains, corresponding to V region function as antigen-binding site, while the other domains have effector functions of Ab. *Fc* crystallizable fragment, *Fab* antigen-binding fragment

heavy chain), IgE (containing ε heavy chain), and IgA (containing α heavy chain) (Fig. 4.1).

Primarily, an Ab functions as a receptor on the surface of the B lymphocytes for specific Ags so that Ag binding to receptor initiates the immune reactions of B lymphocytes for production of an Ab specific for binding an Ag, enabling the successful elimination of Ags by specifically binding to Abs. There are also three effector functions of Abs. The first one is a neutralization function, through which endotoxin shock can be prevented. Abs are able to specifically bind endotoxins such as lipopolysaccharides, which are a cell membrane component of gram-negative bacteria. Otherwise, they bind to their receptor on macrophages evoking endotoxin shock through the induction of a surplus amount of TNF-α and IL-1 produced by macrophages, sometimes resulting in the death of the host. The second function is an opsonization function, through which bacteria can be easily eradicated. Once an Ab binds to the invading bacteria, phagocytes such as macrophages and neutrophils can recognize the bacteria more easily and engulf them, eventually killing them. The last one is complement fixation, through which the bacteria present in the plasma can be eradicated. An Ab activates a complement, which enhances opsonization and lyses some bacteria located in the plasma.

In general, it can be speculated that excessive damages caused by ROS might suppress immune functions including the humoral immunity carried out by the B lymphocytes. However, very few papers have been published regarding this. Only a few papers on the effects of vitamin C on the humoral immune functions have been reported. Liebes L et al. reported that the concentration of ascorbic acid in chronic lymphocytic leukemic (CLL) B cells was much higher than that in normal blood B lymphocytes, while there was no difference between normal and CLL T lymphocytes, indicating that CLL B lymphocytes might need more reducing agents able to react with ROS (Liebes et al. 1981). Our research team has also reported on the effects of vitamin C on B-cell immune responses. An in vitro study showed that vitamin C decreased isotype switching to IgG1 but increased switching to IgG2a. However, an in vivo study showed that vitamin C had no effect on the level of IgG2a but decreased the level of IgG1. Therefore, it is concluded that vitamin C acts indirectly to modulate isotype switching in murine B cells (Woo et al. 2010).

Vitamin C and T Lymphocytes

T lymphocytes have another crucial role in adaptive immune functions, for which the eventual goal is to eliminate foreign substances called Ags. Although simply hidden Ags can be removed by cytotoxic T lymphocytes through the killing of self-cells containing the Ags, exposed Ags can be removed by direct binding of an Ab to an Ag. T lymphocytes have diverse functions and are divided into two subclasses according to their surface markers: CD4 and CD8 lymphocytes. Some CD4 T lymphocytes (helper T lymphocytes; TH) provide B lymphocytes with the crucial signals for the production of an Ab specific for the exposed Ag. They exert direct or indirect influences on several immune responses by secreting numerous soluble mediators such as IL-2, IL-4, IL-5, IL-12, IL-13, IL-17, IL-22, and interferon (IFN)-γ. According to their functions, there are several subtypes of CD4 T lymphocytes (Table. 4.3).

CD8 T lymphocytes have a simply cytotoxic function against the self-cells holding Ags such as the cells infected with viruses and macrophages infected with *Mycobacterium tuberculosis* or *Plasmodium falciparum*. Most cancer cells are transformed genetically and are regarded as altered-self cells so they are considered as foreign and usually killed by CD8 T lymphocytes. These cytotoxic T cells use a very sophisticated action mechanism for killing altered-self cells. All mammalian eukaryotic cells including humans have major histocompatibility gene complex (MHC)-1 molecule, on which Ags are usually loaded as a fragment. Only when the T-cell receptor complex (TCR) on the membrane of the cytotoxic T lymphocytes

Table 4.3 Subtypes of CD4 T lymphocytes

Subtype	Function
TH1	Killing intracellular microbes such as several kinds of virus *M. tuberculosis* (*tuberculosis*) and *P. falciparum* (*malaria*) by activating macrophages through secreting IFN-γ
TH2	Helping B lymphocytes with Ab production through secretion of IL-4, IL-5, and IL-13
TH17	Recruiting neutrophils by activation of fibroblasts through secreting IL-17
TFH	Mediating isotype switching and affinity maturation by providing B lymphocytes in the germinal centers with Ag information
Treg	Downregulating T-lymphocyte activation by inhibiting function of immature dendritic cells

TFH follicular *TH*, *Treg* regulatory *TH*

recognizes the complex of the MHC-1 and Ag fragment on the host cells will the immune function for killing the host cells containing the Ags initiate. This phenomenon is called "MHC restriction."

Several papers studying the effects of vitamin C on T lymphocytes have been reported.

In 1982, Kay et al. studied the effects of vitamin C on T lymphocytes using five hospitalized volunteers. According to their reports, vitamin C deficiency did not alter the T-lymphocyte number or impair the in vitro T-lymphocyte function (Kay et al. 1982). However, unfortunately, they did supplement too small an amount of vitamin C (75 mg/day), and their observation period after the administration of vitamin C was too short (5 weeks).

According to the report of Kennes B and his colleagues, vitamin C enhances the proliferative response of T lymphocytes in vitro as well as tuberculin skin hypersensitivity in vivo in old people (Kennes et al. 1983). Some group of scientists have reported that vitamin C can even inhibit three T-lymphocyte death pathways (spontaneous, growth factor withdrawal, and steroid-induced death pathways), suggesting that vitamin C may function as an immune booster through this mechanism (Campbell et al. 1999). Weeks and Perez have published interesting results showing that the pure form of vitamin C can enhance neurite outgrowth, promote fibroblast adhesion during wound healing, and reduce xenobiotic-induced T-lymphocyte hyperactivation (Weeks and Perez 2007). Grant and his colleagues have reported different aspects of vitamin C. According to their observation of protein expression changes in T lymphocytes caused by vitamin C treatment, vitamin C modulates the T lymphocytes proteome in a time- and dose-dependent manner, and they identified molecular targets for study after antioxidant supplementation in vivo (Grant et al., 2007). Another group of scientists provided some evidence that vitamin C can enhance immune function by showing that vitamin C promotes maturation of murine T lymphocytes, suggesting that the role of vitamin C here might be a cofactor for Jumonji C domain enzymes involved at the level of chromatin demethylation (Manning et al. 2013).

Our research team has kept studying the effects of vitamin C on T lymphocytes. A megadose of vitamin C modulates T-cell functions in Balb/c mice so that a shift in immunity from Th2 to Th1, evident by the elevated IL-2 and gamma interferon and decreased IL-4, takes place only when it is administered during T-cell activation (Noh et al. 2005). In addition, they showed that vitamin C enters mouse T cells as dehydroascorbic acid in vitro via glucose transporter (GLUT)-1 and (GLUT)-3 and does not recapitulate in vivo vitamin C effects, suggesting that, in vivo, vitamin C modulates T lymphocytes indirectly through other components of the microenvironment (Maeng et al. 2009). According to their further study, vitamin C can modulate immune responses by inducing IL-12 secretion of bone marrow-derived dendritic cells, thus stimulating naïve T lymphocytes to differentiate into Th1 cells (Jeong et al. 2011). Our team also revealed that vitamin C could prevent severe liver damages induced by inflammation mediated by IP-10 and pro-inflammatory cytokines such as TNF-α and IFN-γ, of which secretion from hepatic infiltrated T lymphocytes is stimulated by vitamin C deficiency and downregulated by supplementation of vitamin C (Bae et al. 2013).

Vitamin C and NK Cells

Although NK cells belong to the innate immune system, which usually shows a very rapid action for immune responses and a simple action mechanism, their action mechanism is relatively sophisticated. The main function of these cells is killing the host cells (self-cells) with Ags (cells infected with a virus or *M. tuberculosis*) in their cytoplasm or altered-self cells (cancer cells) like CD8 cytotoxic T lymphocytes. However, their action mechanism involved in the killing process is the reverse of that of cytotoxic T lymphocytes. NK cells never kill host cells with the same MHC-1 molecule, but they kill cells with a different MHC-1 molecule or in the absence of a MHC-1 molecule. The role of these two cells, NK cells and cytotoxic T lymphocytes, is mutually complementary in their cytotoxic function.

In the early 1980s, when NK cells were reported for the first time and very few studies existed on them, some scientists reported vitamin C had no effects (Siegel and Morton 1983; Siegel and Morton 1984) on the NK activity or inhibitory effects on the NK activity (Huwyler et al. 1985). Heuser and Vojdani reported that vitamin C could restore the functions of NK, T, and B lymphocytes suppressed by toxic chemicals, where protein kinase had an important role in the induction of the NK activity by vitamin C (Heuse and Vojdani 1997). A new millennium has opened new horizons in the field of study on the effect of vitamin C on the NK activity and function. Atasever and his colleagues showed through an in vitro study that vitamin C could enhance the NK activity significantly in both splenectomized and unsplenectomized β-thalassemia patients, who usually showed a significant decreased NK activity (Atasever et al. 2006). Some Indian scientists published very interesting results showing the effects of vitamin C on the NK activity against tumor cells. They suggest that the role of vitamin C might be in increasing the susceptibility of tumor cells to NK cells by inhibiting platelet aggregation, which offers protection to tumor cells from the cytotoxic activity of the NK cells (Toliopoulos et al. 2011). Our vitamin C research team has published a paper which showed the effects of vitamin C on the NK activity against murine ovarian cancer using Gulo(−/−) mice, which cannot synthesize vitamin C like humans. Gulo(−/−) mice survived a shorter time than that of the normal control or Gulo(−/−) mice supplemented with vitamin C after tumor challenge regardless of treatment with IL-2. NK cells isolated from the Gulo(−/−) mice showed a decreased IFN-γ and decreased mRNA expression of the perforin and granzyme B genes which are strong tools of NK cells for killing cancer cells after they were co-cultured with murine ovarian surface epithelial cells (Kim et al. 2012b). Very recently, some cytotherapy researchers have provided useful information on NK-cell cytotherapy using vitamin C. Huijskens and his colleagues performed an experiment using culture systems applicable for natural killer cell therapy, in which vitamin C enabled NK cells to proliferate faster, resulting in

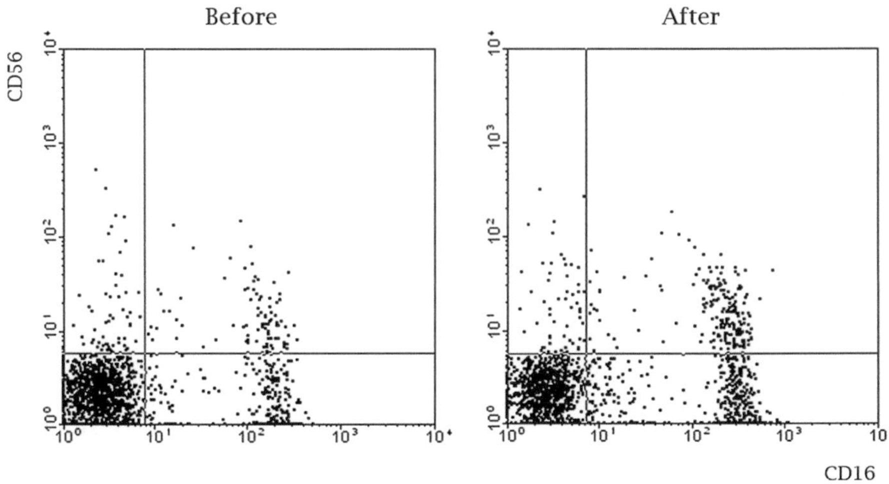

Fig. 4.2 Effect of vitamin C on the number of NK cells. Vitamin C treatment increases the number of NK cells in the peripheral blood. (unpublished our data)

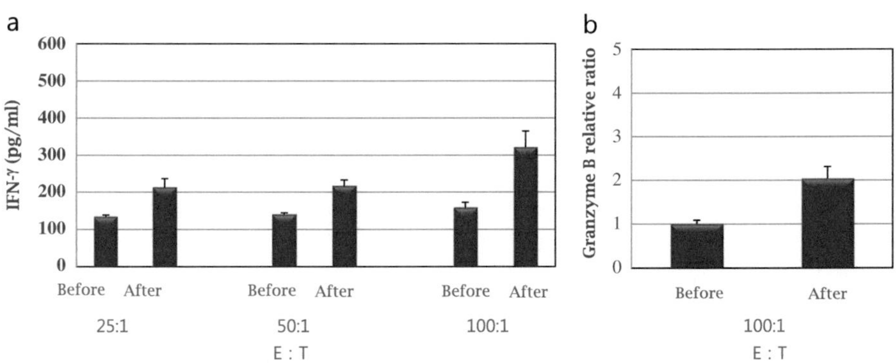

Fig. 4.3 Effects of vitamin C treatment on IFN-γ (**a**) and granzyme B (**b**) production in human NK cells of stomach cancer patients. Vitamin C treatment increased IFN-γ and granzyme B production from human NK cells. *E* T, *effector* (*NK cells*), target (cancer cells). (unpublished our data)

higher cell numbers, which is required for adoptive immunotherapy (Huijskens et al. 2015). Furthermore, our unpublished data show that vitamin C treatment increases the number of NK cells in the human peripheral blood (Fig. 4.2). Additionally, purified NK cells from the peripheral blood of vitamin C-treated humans increase INF-γ and granzyme B production, both of which have key roles in killing cancer cells and virus-infected cells (Fig. 4.3).

Vitamin C and Other Immune Cells Involved in Innate Immunity

In addition to the immune cells described above and involved in specific immune responses, there are some more important immune cells involved in the innate immunity such as macrophages and dendritic cells. Macrophages have a dominant role in eliminating Ags non-specifically with the help of neutrophils recruited by some soluble mediators secreted by macrophages, which is called the inflammatory process, a key process of innate immune responses. During the removal of most of the Ags by macrophages and neutrophils, some Ags are processed by immature dendritic cells, which have the ability of phagocytosis, and then transferred into regional lymph nodes, where the immature dendritic cells are differentiated into mature dendritic cells, which are less phagocytic than immature dendritic cells but have more ability to present Ags to $CD4^+$ T lymphocytes gathered in the regional lymph nodes. Subsequently, Ag-specific CD4 T lymphocytes are generated, and they provide Ag-specific B lymphocytes with a crucial signal for their activation, resulting in the differentiation of Ag-specific B lymphocytes into Ab-producing cells, plasma cells. This process is called the germinal center reaction, which usually takes place in the B-cell area of the peripheral lymphoid organs. Conclusively, macrophages take charge in the primary defense, while dendritic cells are responsible for the linkage of the innate immune reaction to the acquired immunity. Only two cell types have the abilities of phagocytosis and Ag presentation.

In 1985, Ganguly and Waldman reported on the effects of vitamin C deficiency on macrophage functions, especially in aging using guinea pigs, which cannot synthesize vitamin C by themselves. A greater number of aged animals died by week 4 of vitamin C deficiency. Deficient senescent animals had a greater decline in macrophage random migration and bactericidal capacity. Conclusively, their data suggest that vitamin C deficiency may affect macrophage functions in the aged more profoundly and could compromise the parameters of host defenses effective in microbial infections (Ganguly and Waldman 1985). Victor et al. reported on the effects of vitamin C on macrophage functions, especially the phagocytic function of mice with endotoxic shock. Their data suggest that vitamin C can regulate the phagocytic process of macrophages in endotoxic shock, principally decreasing free radical production, and thus, it could reduce the endotoxic shock severity (Victor et al. 2000). The same research team reported the sequential effects of vitamin C on endotoxic shock. According to their studies, endotoxin-induced oxidative stress makes the immune cells such as macrophages and lymphocytes need an appropriate level of antioxidants. Specifically, while lymphocytes take and accumulate vitamin C, macrophages use it (Victor et al. 2001). Vandana et al. (2006) reported on the differential effects of vitamin C on murine macrophages under stress induced by chromium from vitamin E and beta-carotene. According to their reports, the addition of vitamin C inhibited chromium-induced cytotoxicity, ROS generation, and apoptosis, but vitamin E and beta-carotene did not. Giovine's research team studied the effects of vitamin C on chronic pulmonary silicosis caused by murine alveolar macrophages activated by quartz serially. First, vitamin C-pretreated quartz enhances cyclo-oxygenase-2 expression in murine macrophages, thereby suggesting the preventive role of vitamin C from quartz toxicity silicosis (Scarfi et al. 2007). Through their serial experiment, vitamin C-pretreated quartz even stimulates TNF-alpha release in murine macrophages through ROS production and membrane lipid peroxidation, confirming that the prevention of impairment of macrophage phagocytosis by vitamin C might be a way to prevent silicosis induced by quartz exposure (Scarfi et al. 2009). Some scientists have provided an important hint that vitamin C would be helpful for the treatment of nicotine toxicity by providing an immunomodulatory role against nicotine-induced murine peritoneal macrophages in vitro (Mahapatra 2011). Obesity has been known to be related with monocyte infiltration into adipose tissue. Garcia-Diaz et al. (2011) provided some clues that vitamin C can modulate the interaction between adipocytes and macrophages, thereby helping with the management of obesity (Garcia-Diaz et al. 2011). Some Chinese scientists reported that vitamin C mitigates oxidative stress and tumor necrosis factor-alpha in severe community-acquired pneumonia by showing that vitamin C inhibited LPS-induced ROS, DNA damage, TNF-alpha, IL-6, and p38 in macrophage cells (Chen et al. 2014). Recently, the effects of vitamin C on vascular pathophysiology were reported by some European scientists. They observed the effect of vitamin C on tissue factors and factor VII-activating protease expression induced by LPS in human monocyte-derived macrophages and suggested that vitamin C alters pro-inflammatory and procoagulatory processes by inhibition of NF-κB activation and exerts beneficial anti-atherogenic effects on human monocyte-derived macrophages (Parahuleva et al. 2016). Therapeutic strategies against bacterial infection are becoming more important because of the emergence of super bacteria, which are resistant against any kind of antibiotic treatment, even vancomycin, which has been known to be the most potent antibiotic. Recently, the combinational therapy of vitamin C was done along with conventional antibiotics, ofloxacin and chloramphenicol, to kill *S. aureus* by mouse peritoneal macrophages, and it decreased pro-inflammatory cytokines and inflammatory markers like iNOS and COX2 but increased antioxidant enzymes like SOD, catalase, and GSH as well as decreased LPO, thus balancing the ROS and antioxidant status inside the cell. These results may be positively considered as a novel and strong therapeutic strategy against bacterial infection (Dey and Bishayi 2018).

In 2011, Jeong et al. reported that vitamin C can have a role in the Th1 shift of the immune responses, which is mediated by dendritic cells, suggesting a role for vitamin C in modulating immune responses (Jeong et al. 2011). Furthermore, they expanded their boundary of research on the vitamin C effect on dendritic cells to CD8+ memory T cells. They found that vitamin C could enhance the capability of memory T-cell production by mouse bone marrow-derived dendritic cells, suggesting that vitamin C could be considered as a strategy for boosting the DC vaccination potency against tumors (Jeong et al. 2014). Our research team also provided an effective role of vitamin C in enhancing dendritic cell function. We found that a relatively high dose of vitamin C (2.5 mM) induced an increased expression of CD80, CD86, and MHC on the surface of dendritic cells, which was mediated by the activation of p38 MAPK (Kim et al. 2012a).

References

Atasever B, Ertan NZ, Erdem-Kuruca S, Karaka Z. In vitro effects of vitamin C and selenium on NK activity of patients with beta-thalassemia major. Pediatr Hematol Oncol. 2006;23(3):187–97.

Bae S, Cho CH, Kim H, Kim Y, Kim HR, Hwang YI, Yoon JH, Kang JS, Lee WJ. In vivo consequence of vitamin C insufficiency in liver injury: vitamin C ameliolates T-cell mediated acute liver injury in gulo(−/−) mice. Antioxid Redox Signal. 2013;19(17):2040–53.

Campbell JD, Cole M, Bunditrutavorn B, Vella AT. Ascorbic acid is a potent inhibitor of various forms of T cell apoptosis. Cell Immunol. 1999;25(1):1–5.

Chen Y, Luo G, Yuan J, Wang Y, Yang X, Wang X, Li G, Liu Z, Zhong N. Vitamin C mitigates oxidative stress and tumor necrosis factor-alpha in sever community-acquired pneumonia and LPS-induced macrophages. Mediators Inflamm. 2014;2014:426740.

Dey S, Bishayi B. Killing of *S. aureus* in murine peritoneal macrophages by ascorbic acid along with antibiotics Chloramphenicol or Ofloxacin: correlation with inflammation. Microb Pathog. 2018;115:239–50.

Gacia-Diaz DF, Campion J, Quintero P, Milagro FI, Moreno-Aliaga MJ, Martinez JA. Vitamin C modulates the interaction between adipocytes and macrophages. Mol Nutr Food Res. 2011;55(suppl 2):S257–63.

Ganguly R, Waldman RH. Macrophage functions in aging: effects of vitamin C deficiency. Allerg Immunol. 1985;31(1):37–43.

Grant MM, Mistry N, Lunec J, Griffiths HR. Dose-dependent modulation of the T cell proteome by ascorbic acid. Br J Nutr. 2007;97(1):19–26.

Heuse G, Vojdani A. Enhancement of natural killer cell activity and T and B cell function by buffered vitamin C in patients exposed to toxic chemicals: role of protein kinase –C. Immunopharmacol Immunotoxicol. 1997;19(3):291–312.

Huijskens MJ, Walczak M, Sarkar S, Atrfi F, Senden-Gijsbers BL, Tilanus MG, Bos GM, Wieten L, Germeraad WT. Ascorbic acid promotes proliferation of natural killer cell populations in culture systems applicable for natural killer cell therapy. Cytotherapy. 2015;17(5):613–20.

Huwyler T, Hirt A, Morell A. Effect of ascorbic acid on human natural killer cells. Immunol Lett. 1985;10(3–4):173–6.

Jeong YJ, Hong SW, Kim JH, Jin DH, Kang JS, Lee WJ, Hwang YI. Vitamin C-treated murine bone marrow-derived dendritic cells preferentially drive naïve T cells into Th1 cells by increased IL-12 secretions. Cell Immunol. 2011;266(2):192–9.

Jeong YJ, Kim JH, Hong JM, Kang JS, Kim HR, Lee WJ, Hwang YI. Vitamin C treatment of mouse bone marrow-derived dendritic cells enhanced CD8(+) memory T cell production capacity of these cells in vivo. Immunobiology. 2014;219(7):554–64.

Kay NE, Holloway DE, Hutto SW, Bone ND, Duane WC. Human T-cell function in experimental ascorbic acid deficiency and spontaneous scurvy. Am J Clin Nutr. 1982;36(1):127–30.

Kennes B, Dumont I, Brohee D, Hubert C, Neve P. Effect of vitamin C supplements on cell-mediated immunity in old people. Gerontology. 1983;29(5):305–10.

Kim HW, Cho SI, Bae S, Kim H, Kim Y, Kang JS, Lee WJ. Vitamin C up-regulates expression of CD80, CD86 and MHC class II on dendritic cell line, DC-1 via the activation of p38MAPK. Immune Netw. 2012a;12(6):277–83.

Kim JE, Cho HS, Yang HS, Jung DJ, Hong SW, Hung CF, Lee WJ, Kim D. Depletion of ascorbic acid impairs NK cell activity against ovarian cancer in a mouse model. Immunobiology. 2012b;217(9):873–81.

Liebes L, Krigel R, Kuo S, Nevria D, Pelle E, Silber R. Increased ascorbic acid in chronic lymphocytic leukemia B lymphocytes. Proc Natl Acd Sci USA. 1981;78(10):6481–4.

Maeng HG, Lim H, Jeong YJ, Woo A, Kang JS, Lee WJ, Hwang YI. Vitamin C enters mouse T cells as dehydroascorbic acid in vitro and does not recapitulate in vivo vitamin C effects. Immunobiology. 2009;214(4):311–20.

Mahapatra SK, Chakraborty SP, Roy S. Immunomodulatory role of *Ocimum gratissimum* and ascorbic acid against nicotine-induced murine peritoneal macrophages in vitro. Oxidative Med Cell Longev. 2011.

Manning J, Mitchell B, Appradurai DA, Shakya A, Pierce LJ, Wang H, Nganga V, Swanson PC, Tantin D, Spangrude GJ. Vitamin C promotes maturation of T-cells. Antioxid Redox Signal. 2013;10(17):2054–67.

Noh K, Lim H, Moon SK, Kang JS, Lee WJ, Lee D, Hwang YI. Mega-dose of vitamin C modulates T cell functions in Balb/c mice only when administered during T cell activation. Immunol Lett., 15. 2005;98(1):63–72.

Parahuleva MS, Jung J, Burgazli M, Erdogan A, Parviz B, Holscherman H. Vitamin C suppresses liposaccharide-induced procoagulant response of human monocyte-derived macrophages. Eur Rev Med Pharmacol Sci. 2016;20(10):2174–82.

Scarfi S, Benatti U, Pozzoloini M, Clavarino E, Ferraris C, Magnone M, Valiano L, Giovine M. Ascorbic acid-pretreated quartz enhances cyclo-oxygenase-2 expression in RAW 264.7 murine macrophages. FEBS J. 2007;274(1):60–73.

Scarfi S, Magnone M, Ferraris C, Pozzoloini M, Benvenuto F, Benatti U, Giovine M. Ascorbic acid pre-treated quartz stimulates TNF-alpha release in RAW 264.7 murine macrophages through ROS production and membrane lipid peroxidation. Respir Res. 2009;19:20–5.

Siegel BV, Morton JI. Vitamin C and immunity: natural killer (NK) cell factor. Int J Vitam Nutr Res. 1983;53(2):179–83.

Siegel BV, Morton JI. Vitamin C and immunity: influence of ascorbate on prostaglandin E2 synthesis and implications for natural killer cell activity. Int J Vitam Nutr Res. 1984;54(4):339–42.

Toliopoulos IK, Simos YV, Daskalou TA, Verginadis II, Evangelou AM, Karkabounas SC. Inhibition of platelet aggregation and immunomodulation of Nk lymphocytes by administration of ascorbic acid. Indian J Exp Biol. 2011;49(12):904–8.

Vandana S, Ram S, Ilavazhagan M, Kumar GD, Banerjee PK. Comparative cytoprotective activity of vitamin C, E and beta-carotene against chromium induced oxidative stress in murine macrophages. Biomed Pharmacother. 2006;60(2):71–6.

Victor VV, Guayerbas N, Puerto M, Medina S, De la Fuente M. Ascorbic acid modulates in vitro the function of macrophages from mice with endotoxic shock. Immunopharmacology. 2000;46(1):89–101.

Victor VM, Guayerbas N, Puerto M, De la Fuente M. Changes in the ascorbic acid levels of peritoneal lymphocytes and macrophages with endotoxin-induced oxidative stress. Free Radic Res. 2001;35(6):907–16.

Weeks BS, Perez PP. A novel vitamin C preparation enhances neurite formation and fibroblast adhesion and reduces xenobiotic-induced T-cell hyperactivation. Med Sci Monit. 2007;13(3):BR51–8.

Woo A, Kim JH, Jeong YJ, Maeng HG, Lee YT, Kang JS, Lee WJ, Hwang YI. Vitamin C act indirectly to modulate isotype switching in mouse B cells. Anat Cell Biol. 2010;43(1):25–35.

Chapter 5
Common Cold and Flu

The common cold is a very common viral infectious disease, as its name indicates, which no one can avoid during their whole life. Medically, the mortality rate of this disease has been known to be zero. Even flu, which is more morbid and toxic than the common cold, has a mortality rate near around zero. Most of the causative agents of the common cold are viruses, especially RNA viruses, such as the rhinovirus, corona virus, adenovirus, Coxsackie virus, and so on. Even though there are some common colds of bacterial origin, actually, their proportion is very small. In addition, they can be treated successfully by potent antibiotics, whereas optimal regimens for the successful eradication of the common cold of viral origin are not available yet. The main symptoms of the common cold, although they depend upon the initial site of infection, are runny nose, mild fever, throat pain, cough, and headache (rare). Because the common cold is a disease occurring only in the respiratory system, especially in the upper respiratory system, physicians usually call it a URI (upper respiratory infection). Therefore, before explaining the common cold in detail, the structures of the respiratory system will be reviewed briefly.

Structures of the Respiratory System

Respiration can be defined as gas exchange between O_2 and CO_2, which takes place in the lungs (pulmonary respiration, in which O_2 moves from air space to capillaries, while CO_2 moves in the opposite direction) as well as in tissues (tissue respiration, in which O_2 moves from blood to cells, while CO_2 moves in the opposite direction). The final goal of respiration is to supply all cells in the body with O_2, one of the important components for power generation, and to expel out CO_2 generated in the cells during the power generation process. The respiratory system is a group of structures for pulmonary respiration, consisting of an air conduction portion and a respiratory portion (Fig. 5.1).

W. J. Lee, *Vitamin C in Human Health and Disease*,
https://doi.org/10.1007/978-94-024-1713-5_5

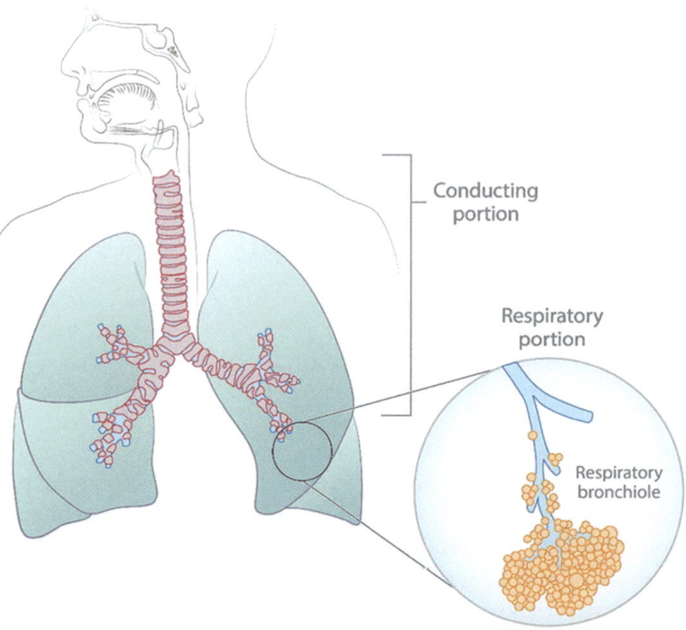

Fig. 5.1 Respiratory system consists of conducting and respiratory portion. The nose, pharynx, larynx, trachea, bronchus, and several divisions of bronchioles to terminal bronchiole are included in conducting portion. Respiratory portion, where the gas exchanges take place, contains respiratory bronchiole and alveoli

The air conduction portion is composed of the nasal cavity, pharynx, larynx, trachea, bronchus (divided in a dichotomous manner more than 20 times), and finally terminal bronchiole in that order from outside to inside, where gas exchange never takes place, but gases (O_2 or CO_2) are transported from the outside to inside or vice versa. The main functions of the air conduction portion are conditioning the air for pulmonary respiration as well as air conduction as its basic function. Conditioning of the air includes three actions, (1) air cleaning, (2) air warming, and (3) air moistening. Interestingly, the nasal cavity, the first part of the conduction portion, has complicated structures for conditioning of the air such as dome-shaped nasal alae (their inner space is the nasal ventricle), making air warm and moistened by slowing the air speed, and three shelves (inferior, middle, and superior conchae) projected from the nasal septum in the nasal cavity, making air warm, moistened, and clean by increasing the contact of the ingested air with typical respiratory mucosal epithelium having moisture and warmth from the rich capillaries by causing air turbulence, in addition to its basal function, air conduction. Especially, mucin secreted from the typical respiratory epithelium of the nasal cavity is so sticky that it gets rid of the particulate materials such as dusts and microorganisms easily through their adhesion to the sticky mucin, which is called the "air cleaning process." The larynx also has its proper function, phonation, for which it has several small cartilages and small muscles connecting each cartilage, besides its air conduction function. Needless to say, the air passing through the entire air conduction portion is fully conditioned air, e.g., cleaned, moistened, and warmed appropriately, arriving at the respiratory portion. The air conduction portion from the bronchus to the terminal bronchiole is called the bronchial tree, which also has a typical respiratory epithelium on their luminal surface. The typical respiratory epithelium consists of several cell types: (1) tall columnar cells are the most abundant ones and have numerous cilia on their apical surface; (2) mucous goblet cells, looking like a typical wine glass, have many polysaccharide-rich mucin droplets on their surface; and (3) there are some miscellaneous cells.

Functionally, all cilia of the tall columnar cells are fully covered by spreading out the polysaccharide-rich mucous droplets produced by mucous goblet cells, forming the mucociliary blanket. The sticky mucin component laid on the cilia adsorb chemical substances, tiny particles, and even microorganisms, which are removed effectively by one-way movement (from inside to outside) of the mucociliary blanket. Therefore, a substantial amount of mucous fluid drains from the bronchial tree into esophagus, a pathway to the GI tract that is the crossroads in the laryngopharyngeal area between the respiratory and GI tracts, every day in humans.

The respiratory portion starts from the last and smallest bronchiole, called respiratory bronchioles, which have alveoli intermittently on their way to the terminal portion of respiratory pathway, the alveolar sacs, which are composed of numerous alveoli. Wide networks of capillaries cover the external surface of each alveolus, forming the blood air barrier, consisting of the alveolar epithelium and endothelium of its covering capillary, and the fused basal laminae present between them (Fig. 5.2).

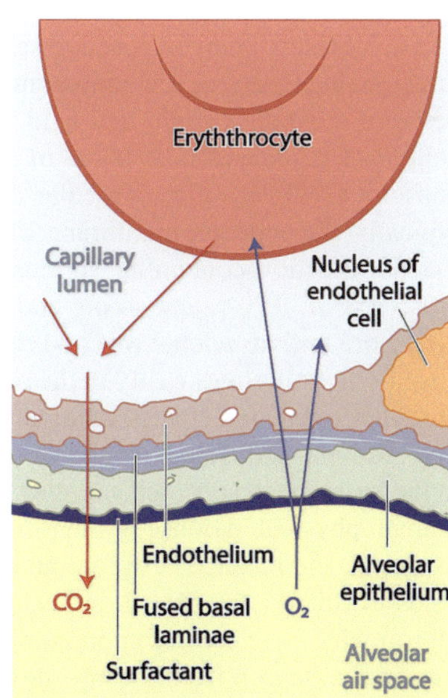

Fig. 5.2 Blood-air barrier. O_2 (blue arrows) in the air inhaled into alveoli gets through the blood-air barrier into the capillary, while CO_2 (red arrows) transported from cells to the alveolar capillary is penetrated to the alveolar space through the blood air barrier, which consists of alveolar epithelium, fused basal lamina, and capillary endothelium

Conclusively, O_2 in the alveoli moves into blood capillaries through the blood air barrier, while CO_2 in the blood capillaries is transported to the air spaces in the alveoli through the blood air barrier, eventually evacuated to the outside of the body.

Pathogenesis of the Common Cold

There is no doubt in the fact that the direct initiating factor of the common cold is low temperatures, i.e., coldness, especially relative coldness, which means a sense of feeling cold due to an unexpected sudden change in the temperature from high to low. Here, coldness does not mean an absolute extremely low temperature routinely experienced in both arctic areas year-round. Therefore, although the outdoor temperature can be higher than 30 °C, people attending a meeting in a room where the air-cooling system is operating well with the indoor temperature kept around 24–25 °C cannot help but feel cold for a relatively long time, and they can catch the common cold. Generally speaking, traditionally, the incidence rate of the common cold during the hot summer has been very low. In the Korean proverb, "Even the dogs do not catch colds in May and June of the lunar calendar." Understanding that May and June in the lunar calendar indicate a midsummer time, it means that Korean people rarely catch cold in the midsummer time. However, nowadays, the rate has become relatively high such that it is very common to see cold patients during the summer time because air-cooling systems are found everywhere. It is no wonder to see that the incidence rate of the common cold is not particularly high in the arctic areas where absolute coldness persists year-round, as well as during the cold winter time, when severe cold temperatures persist, because they do not usually feel cold due to routinely being exposed to severe coldness as wells as having low viral colonization due to severe low temperatures. Actually, relative coldness means the temperature at which people feel cold in real life. When people feel cold or chilly, a couple of responses take place in their body to keep the core body temperature constant (37.5 °C). First, vascular constriction around the inlet area of air occurs, typically the mucous membrane of the nasal cavity, resulting in a reduction of mucus secretion containing various antiviral soluble mediators, such as interleukins, TNF-α, IFN-γ, and so on, and followed by an explosive increase in the nasal viral flora such as adenovirus and rhinovirus by their rapid proliferation. The nasal viral flora proliferates easily in the relatively cold weather, which causes the relative humidity to become low. A relatively low temperature and dryness are good conditions for viral proliferation and invasion to the body. Dryness usually causes injuries in the nasal mucosa, which supplies the route for the initial viral invasion. When the human physical defense system weakens or is broken and viral proliferation increases, the balance between the human body defense system and the viral flora tends to shift to the viral flora, resulting in catching the common cold. That is the reason why there are common cold or flu outbreaks in the late fall season and the transitional season from the hot summer to the cold winter. When the common cold starts just after a prolonged exposure to relative coldness, the first symptom of the common cold is usually a runny nose, which functions to wash out the increased viral flora so that the number of virus infected cells decreases. Moreover, people in good physical condition ordinarily do not catch the cold, while people who are not healthy ordinarily due to prolonged fatigue, overwork, or insomnia usually catch the cold. Once people catch the cold, the next symptoms after a runny nose are a sore throat and systemic fever. A sore throat means that tissue invasion by the proliferated viral flora has started, followed by inflammation which usually consists of four classical signs of redness, swelling, pain, and heat. Systemic fever is a very important innate defense. The higher the body temperature is to some extent, the lower the proliferation rate of the virus is. Therefore, normal body temperature itself (37.5 °C) is a very important body defense against bacterial or viral infection. Actually, bacterial or viral proliferation is strongly inhibited by high temperatures (higher than 37.5 °C). Because it takes a little time (minimally 2–3 days) to activate the adaptive, specific immunity for invading bacteria or a virus, the body defense system uses the high fever as a non-specific defense tool immediately upon viral infection, which works in a very short time. That is the reason why early use of antipyretic agents should not be done for a high fever occurring in the common cold or flu. However, a high fever should be cautiously controlled by pediatricians in the case of young-aged people in whom brain damage is likely to be caused by febrile convulsions. Acute symptoms of the common cold are usually terminated within 3 days, at the longest 5–7 days, before the specific adaptive immune system is fully active, unless a secondary bacterial infection is also present. There are a couple of reasons why actually vaccines for the common cold are not available. One is that the common cold comes from an imbalance between the human defense system and the normal viral flora, which normally reside at the initiation part of the respiratory system such as the nasal cavity and nasal or oral pharynx so that they usually do not induce new immunity against the normal floral virus. The second reason is because there are more than 200 viruses that cause the common cold. The third reason is because the common cold has been called a self-limited disease in which it takes just a few days to be cured so that mainly the innate immunity must work but the adaptive immunity does not need to be activated for the eventual termination of the common cold.

One more thing is to be considered seriously; the common cold and flu are totally different. The flu is never a severe cold. Clinically, it is not easy to tell the flu from a common cold, although there have been some differences in clinical symptoms. Briefly, the main symptoms of the common cold are runny nose, mild fever, cough, and sometimes sore throat, while those of the flu are high fever, myalgia (muscle pain), headache, sore throat, and generalized weakness (lethargy) (Table. 5.1).

Unlike the common cold, which is usually not infected by a foreign pathogen but is just a kind of battle between the body defense system and the normal flora of viruses residing in the upper respiratory system, the flu is a disease caused by an influenza virus infection. Thus, the common cold takes place at any time year-round whenever people feel coldness for a relatively long time, while flue outbreaks occur only during some restricted periods (in Korea from October to April of the next year) when the influenza virus is prevalent.

Table 5.1 Differences in clinical symptoms between the flu and the common cold

	The flu	The common cold
Runny nose	Yes (less common)	Yes (very common)
Cough	Yes (less common)	Yes (common)
Sore throat	Yes	Yes
High fever	Yes	Rare (low-grade fever)
Headache	Yes	Rare
Myalgia	Yes	Rare
Lethargia	Yes	Rare

Change in the Common Cold or the Flu

In March 2003, the WHO reported officially that a new unidentified fatal disease has appeared in Hong Kong. It was discovered that the disease is a kind of variant of the common cold caused by a variant coronavirus, which supposedly originated from musk cats raised by some farmers in Guangdong, China, in November 2002. The coronavirus is a single-stranded RNA virus and has a species specificity as the causative viral agent of infection. However, a genetically modified coronavirus of the musk cat has a new capability to infect beyond the interspecies barrier. It was not just the matter of crossing the interspecies barrier. Even though its genetic structure has changed, it is just one of the new causative agents of the common cold, which causes people to become infected with the common cold. Unfortunately, some of the patients infected by this variant coronavirus have died. The autopsy of the dead cases revealed that the specific cause of death was viral pneumonia, which rarely occurs in routine common cold patients. Since then, the common cold caused by the variant coronavirus is called severe acute respiratory syndrome (SARS). Epidemiological data show that the mortality rate of SARS has reached around 10% (about 780 people dead out of about 83,000 patients), which is high enough to cause people to panic, considering that the mortality rate of the original common cold is near zero.

At the end of March 2009, a new strain of type A influenza virus was reported for the first time from a 10-year-old child admitted to one clinic located in San Diego, USA, with the chief complaints of high fever, cough, and vomiting. It turned out to be a novel swine-origin influenza A (H1N1) virus that caused the infection. Thereafter, this novel flu has spread out to the whole world, resulting in the occurrence of more than 130,000 patients and dead cases of more than 800 people. Fortunately, this novel flu could be controlled successfully by the administration of Tamiflu, which does not have any treatment effects on the seasonal flu despite having the same serotype (H1N1). Therefore, many researchers think that the actual mortality rate of this novel swine-origin influenza A might be more than 10%. Influenza virus has two kinds of antigens, which consist of hemagglutinin (H) and neuraminidase (N). Hemagglutinin has 15 subtypes (H1~H15), while neuraminidase has 9 subtypes (N1~N9). A combination of H subtypes and N subtypes makes theoretically 135 sub-serotypes of influenza A virus such as H1N1, H3N2, H5N1, H7N9, and H15N9. According to scientific reports, humans are usually infected by influenza virus with low numbers of H (H1~H3), while avian are usually infected by influenza virus with high numbers of H (H5~H15). The H protein is known to be used for viral attachment to the host cell, after which the virus proliferates using the replication machinery of the host cell. The H protein has high species specificity. Unfortunately, this principle has been broken, welcoming a new millennium, in 2000. In late 2003, shocking news was reported that some people living in East Asia had been infected by a highly pathogenic avian influenza (AI), for which their serotype was H5N1, which had never been reported before in human medical history at that time. Even worse, this H5N1 avian influenza showed an extremely high mortality rate (about 70%), even though its infectivity is low. However, there is good fortune in this misfortune because infection cases between humans have not been identified yet. Recently (spring 2013), a new serotype of AI has appeared in some areas of China, for which the serotype is H7N9. According to some reports presented at the meeting of "Options for the Control of Influenza" held in Cape Town from September 5 to 9, 2013, this new AI is expected to have a higher infectivity and toxicity than H5N1 AI, and in the fall of 2013, a pandemic of this new AI was forecast by the Centers for Disease Control and Prevention (CDC) of the USA.

Vitamin C Against the Common Cold

In Korea, for the last 20 years, the number of people taking a couple of grams of vitamin C daily has increased steadily such that in 2017, it was presumed that about 15 million people were taking a megadose of vitamin C daily. The most frequent cases, which I have observed by telephone or e-mail interviews, have been many people taking a megadose of vitamin C for more than 4–5 years, and for a couple of years, they have never caught a cold, which they used to suffer from one to two times a year when they did not take a megadose of vitamin C. Twelve years ago, I was asked to meet a CEO who was in his mid-50s and suffered from multiple episodes of a cold yearly for more than 30 years. He was admitted regularly to a hospital for 1 month due to pulmonary complications from the cold. Immediately after meeting with me, he started to take a megadose of vitamin C daily. Since then, he has never caught a cold for more than 10 years. This longitudinal study has very meaningful value in estimating the effects of vitamin C on the common cold, even though it is only a single case.

Clinical studies on the effects of vitamin C on colds presented by some clinicians have shown generally negative results. Regular vitamin C consumption may decrease the duration of cold symptoms but does not affect the symptom severity or act as a prophylaxis (Heimer et al. 2009). The failure of vitamin C supplementation to reduce the incidence of colds in a normal population indicates that preventing a cold by taking routine megadoses of vitamin C is not rationally justified for community use. However, evidence suggests that it could be justified in people exposed to brief periods of severe physical exercise or cold environments (Douglas et al. 2007). Their conclusion is that vitamin C does not have a curative or preventive effect on colds but shortens the duration of the illness. The common cold is a self-limited viral disease. For a curative treatment, the optimal dose of vitamin C and the right method of administration should be considered. According to a pharmacokinetic study of vitamin C after oral administration, it is most desirable to take vitamin C every 6 h because vitamin C has a 6-h cycle of metabolism as mentioned in Chap. 1, the Introduction. Actually, when it is orally administered, it reaches peak concentration in the peripheral blood 3 h after the administration, and it comes back to the basal level of concentration 6 h after the administration. However, almost all clinical experiments on the effects of vitamin C published so far have been performed on a single dose of vitamin C. Furthermore, the daily vitamin C dose used in the above published papers was less than 1000 mg. In addition, it is actually impossible to perform clinical experiments that can confirm if vitamin C can cure or prevent a cold because a cold has a short period as a self-limited disease. Once cold patients visit the clinic for treatment, the number of viruses is too high to be controlled by any potent medication such that the balance between the host and the virus is shifted to the virus and the patient usually becomes severely ill. During such an active phase of the cold, no drug can stop the disease process of a cold from occurring in the upper respiratory tract. Vitamin C can prevent the onset of a cold if people follow a strict regimen taking a megadose of vitamin C three times daily because it can inhibit the initial phase of viral proliferation, when the number of viruses is not so high and still controllable by the host defense system (Furuya et al. 2008). That is the case with Tamiflu, which has been used for successful prevention of the swine flu (H1N1). The correct method of administering Tamiflu is to take it within 48 h after the occurrence of fever. If it is taken 48 h or later after a fever occurs, it has no effect on the treatment of swine flu. Considering the operating mechanism of Tamiflu, which binds to the hemagglutinin protein of the influenza virus and inhibits viral attachment to the host cell resulting in failure of viral replication using the replication machinery of the host cell, Tamiflu does not cure the cold but prevents it because Tamiflu blocks the initial proliferation of the influenza virus.

In Vivo Experiment Showing the Effects of Vitamin C on the Flu

There have been some reports on the effects of vitamin C on the flu (Li et al. 2006; Sladkova and Kostolansky 2006; Banerjee and Kaul 2010). To observe the effects of vitamin C on flu, an in vivo experiment using Gulo(−/−) mice, which are unable to synthesize vitamin C like humans because the gene encoding L-gulono-γ-lactone oxidase is knocked out artificially, has been performed by my research team. The experimental animals were divided into three groups with six mice in each group: Group 1, wild-type, normal mice; Group 2, Gulo(−/−) mice without supplementation of vitamin C; and Group 3, Gulo(−/−) mice with supplementation of oral vitamin C. The mice in each group were infected with H3N2 influenza virus, and then,

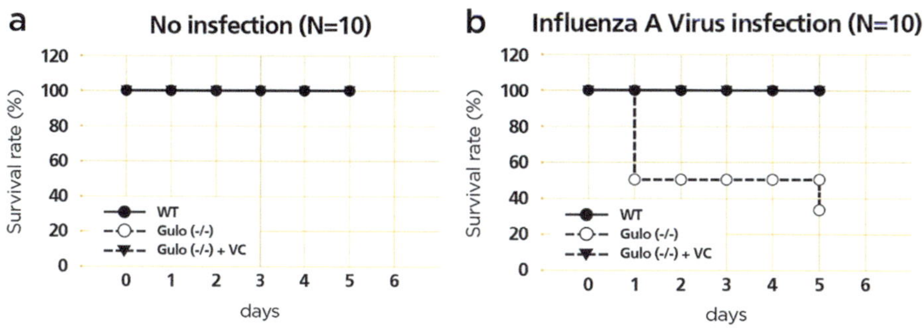

Fig. 5.3 Difference of survival rates among each groups. Only vitamin C-deficient mice (Gulo(−/−)) showed decreased survival rates by influenza infection (**b**), while no dead mice were observed in any groups without influenza infection (**a**). WT, normal mice group; Gulo(−/−), mice group which cannot produce vitamin C like humans; Gulo(−/−) + VC, Gulo(−/−) mice group fed with daily vitamin C in the drinking water. (Kim et al. 2013)

Fig. 5.4 Viral titration in the lungs from H3N2 influenza virus-infected mice using RT-PCR. The highest viral titer was observed in the lung of vitamin-deficient mice (Gulo(−/−)). WT, normal mice group; Gulo(−/−), mice group which cannot produce vitamin C like humans; Gulo(−/−) + VC, Gulo(−/−) mice group fed with daily vitamin C in the drinking water. (Kim et al. 2013)

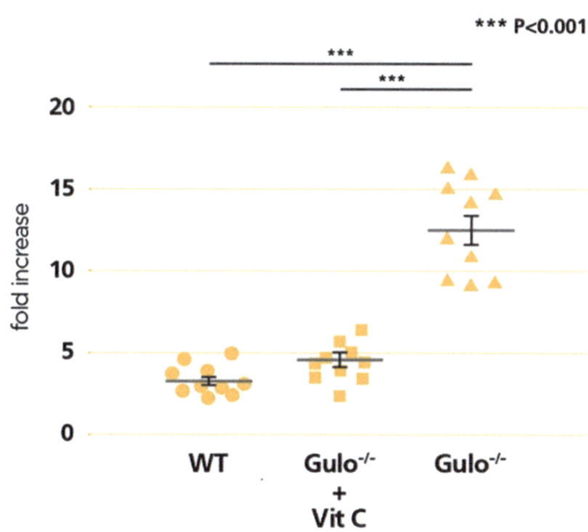

the survival rate was checked. All the mice in the three groups showed symptoms of the flu such as fever, rhinorrhea, lethargy, and so on. Interestingly, 5 days after infection, 80% of the mice in Group 2 were dead, while no mice in Groups 1 and 3 died (Fig. 5.3).

The autopsies showed that the lung tissues from the dead mice were loaded with so much influenza virus, indicating that the mice in Group 2 died of viral pneumonia. However, the lung tissues from the mice in Groups 1 and 3, which did not die from the infection, did not have the influenza virus (Fig. 5.4) (Kim et al. 2013).

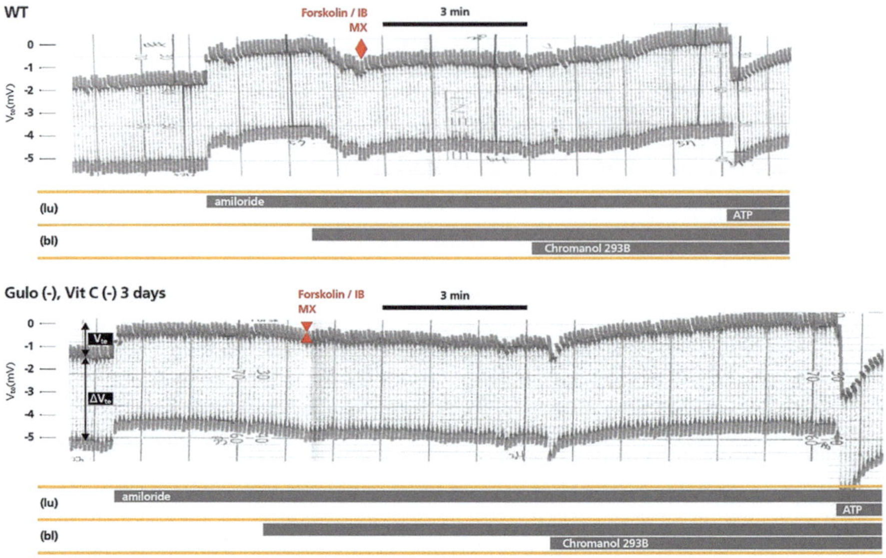

Fig. 5.5 Transepithelial voltage (Vte) and short-circuit current (Isc) analysis. Vitamin C deficiency causes inhibition of cAMP-dependent secretion of chloride in the mouse airway epithelium so that the epithelium does not maintain fluidity on their surface, resulting in facilitation of viral invasion and proliferation

These results strongly suggest that vitamin C in a living body might prevent influenza virus from spreading down from the upper respiratory tract to the lower respiratory system such as the bronchi, bronchioles, and alveoli, even though the mechanism involved in the preventing process has not been elucidated yet. However, vitamin C deficiency causes inhibition of cAMP-dependent secretion of chloride in the mouse airway epithelium so that the epithelium does not maintain fluidity on their surface (our unpublished data) (Fig. 5.5).

The dryness of the mucosal epithelium facilitates viral proliferation and invasion. Furthermore, vitamin C deficiency dramatically decreases SVCT expression on the mucosal epithelium of respiratory tracts, leading to an abolition of the prevention effects of vitamin C from flu infection (our unpublished data) (Fig. 5.6).

In addition, levels of interferons (IFNs) in the bronchoalveolar lavage (BAL) fluid (Fig. 5.7), which have very critical roles in the defense of viral infection, are markedly low in vitamin C insufficiency group. However, levels of representative pro-inflammatory cytokines, such as TNF-α and IL-1α and IL-1β in BAL fluid, are

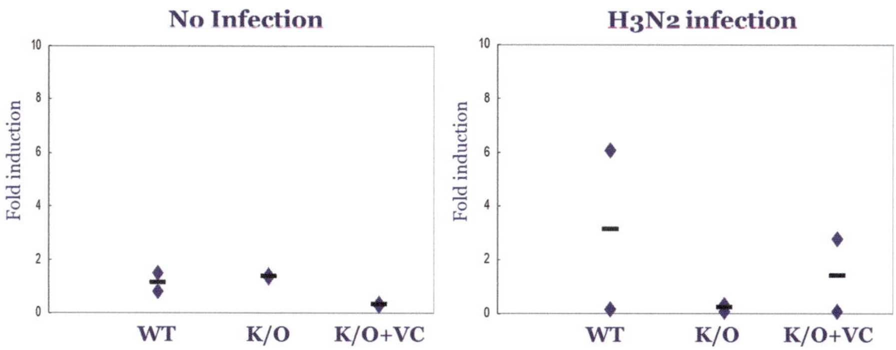

Fig. 5.6 SVCT2 expression in lungs from H3N2 influenza virus-infected mice using RT-PCR. Influenza infection induced increased expression of SVCT2 in the lungs of WT and K/O+VC groups, while it did not increase SVCT2 expression in the lungs of the vitamin C-deficient mice (K/O group). WT, normal mice group; K/O, mice group which cannot produce vitamin C like humans; K/O + VC, K/O mice group fed with daily vitamin C in the drinking water

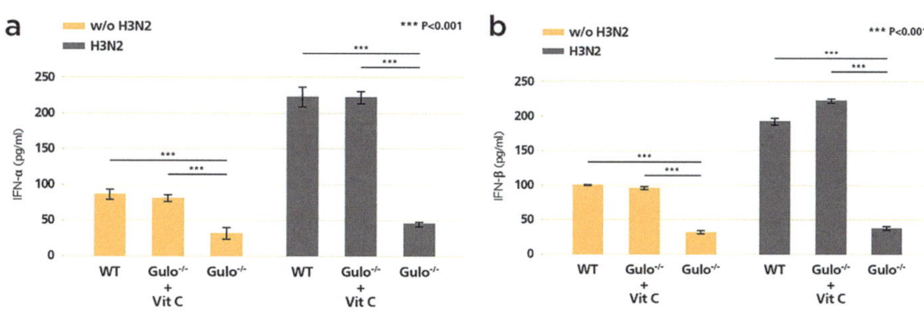

Fig. 5.7 Cytokine production in bronchoalveolar lavage (BAL) fluid (1). Vitamin C deficiency decreases production of IFN-α (**a**) as well as IFN-β (**b**), more significantly during H3N2 influenza infection. WT, normal mice group; Gulo(−/−), mice group which cannot produce vitamin C like humans; Gulo(−/−) + VC, Gulo(−/−) mice group fed with daily vitamin C in the drinking water

significantly high in vitamin C insufficiency group. Inflammatory cells are also markedly increased in BAL fluid (Fig. 5.8) (Kim et al. 2013).

These all in vivo data provide the immunological evidences that daily supplementation of high-dose vitamin C can prevent influenza infection and that, in extreme case, it can save the life.

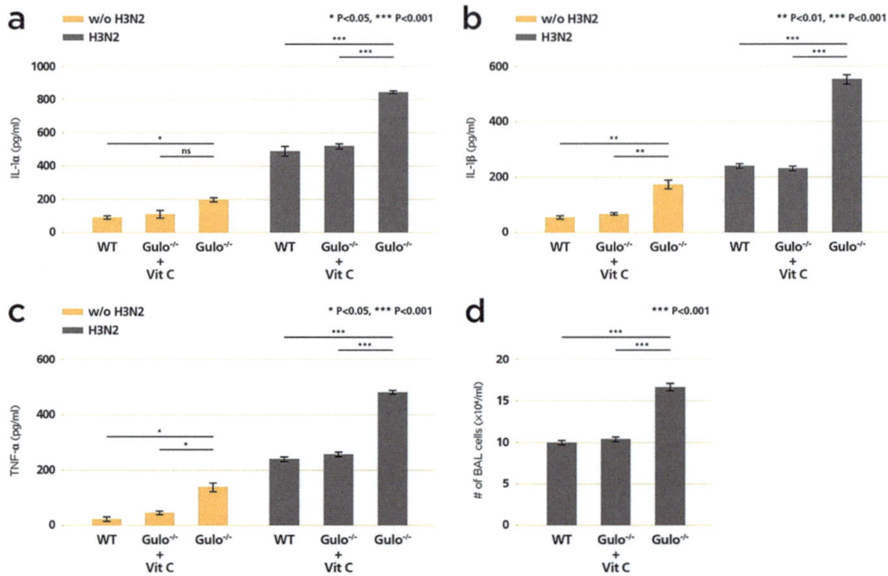

Fig. 5.8 Cytokine production in bronchoalveolar lavage (BAL) fluid (2). Vitamin C deficiency increases production of several pro-inflammatory cytokines such as IL-1α (**a**), IL-1β (**b**), and TNF-α (**c**), especially during H3N2 influenza infection, and it also causes increase of the inflammatory cells in BAL fluid (**d**). WT, normal mice group; Gulo(−/−), mice group which cannot produce vitamin C like humans; Gulo(−/−) + VC, Gulo(−/−) mice group fed with daily vitamin C in the drinking water

References

Banerjee D, Kaul D. Combined inhalational and oral supplementation of ascorbic acid may prevent influenza pandemic emergency: a hypothesis. Nutrition. 2010;26(1):128–32.

Douglas RM, Hemila H, Chalker E, Treacy B. Vitamin C for preventing and treating the common cold. Cochrane Database Syst Rev. 2007;18(3):CD000980.

Furuya A, Uozaki M, Yamasaki H, Arakawa T, Arita M, Koyama AH. Antiviral effects of ascorbic and dehydroascorbic acids in vitro. Int J Mol Med. 2008;22(4):541–5.

Heimer KA, Hart AM, Martin LG, Rubio-Wallace S. Examining the evidence for the use of vitamin C in the prophylaxis and treatment of the common cold. J Am Acad Nurse Pract. 2009;21(5):295–300.

Kim Y, Kim H, Bae S, Choi J, Lim SY, Lee N, Kong JM, Hwang Y, Kang JS, Lee WJ. Vitamin C is an essential factor on the anti-viral immune responses through the production of interferon--α/β at the initial stage of influenza A virus (H3N2) infection. Immune Netw. 2013;13(2):70–4.

Li W, Maeda N, Beck MA. Vitamin C deficiency increases the lung pathology of influenza viruainfected gulo(−/−) mice. J Nutr. 2006;136(10):2611–6.

Sladkova T, Kostolansky F. The role of cytokines in the immune response to influenza A virus infection. Acta Virol. 2006;50(3):151–62.

Chapter 6
Central Nervous System (CNS)

The CNS is like a control tower watching over the whole body. Functionally, it consists of a sensory and motor system, and both usually collaborate with each other for proper functions. Structurally, the CNS consists of the brain located in the cranial cavity and the spinal cord located in the vertebral canal. The brain is composed of the cerebrum (forebrain), the diencephalon (thalamus-related structures), the brain stem (midbrain, pons, and medulla oblongata), and the cerebellum. The spinal cord is divided into the cervical, thoracic, lumbar, sacral, and coccygeal segments (Fig. 6.1).

There are several different types of cells in the CNS: neurons are typical cells transmitting information between brain regions and have key roles in the CNS, and there are neuroglial cells, which are smaller than neurons but are present at 10–50-folds higher than neurons. They are divided into two groups: one is the central ones which consist of microglia cells taking part in the defense of the CNS, astrocytes and oligodendrocytes performing supportive functions, and ependymal cells covering the inner surface of the brain ventricles. The other is the peripheral ones, which consist of Schwann cells and satellite cells. Neurons usually have a unique morphology compared to other somatic cells, which usually only have a cell body. They consist of a cell body and a long projection(s) so that the length of some neurons reaches a couple of 10 cm (Fig. 6.2).

This morphological uniqueness reflects of how the neurons are working. Basically, the CNS is a system for the neuronal transmission of some information, which is sensory- or motor-related. Therefore, the patterns of neuronal transmission usually involve a two-cell relay system. For example, a touch sense from the skin is transmitted to some neuron (first neuron) in the spinal cord or the brain stem via the nerve fibers, which are the projection fibers (called the peripheral sensory nerves) of the neuron in the spinal cord. And then, the signals for the touch sense from the skin are transferred to the long projection fibers (called the tracts) of the second neuron located in the cerebrum of the brain, which is called synapsis. Eventually the signal is accepted and interpreted in the second neuron located in the cerebral cortex. For the motor impulse, the way of neural transmission is reversed so that the motor

W. J. Lee, *Vitamin C in Human Health and Disease*,
https://doi.org/10.1007/978-94-024-1713-5_6

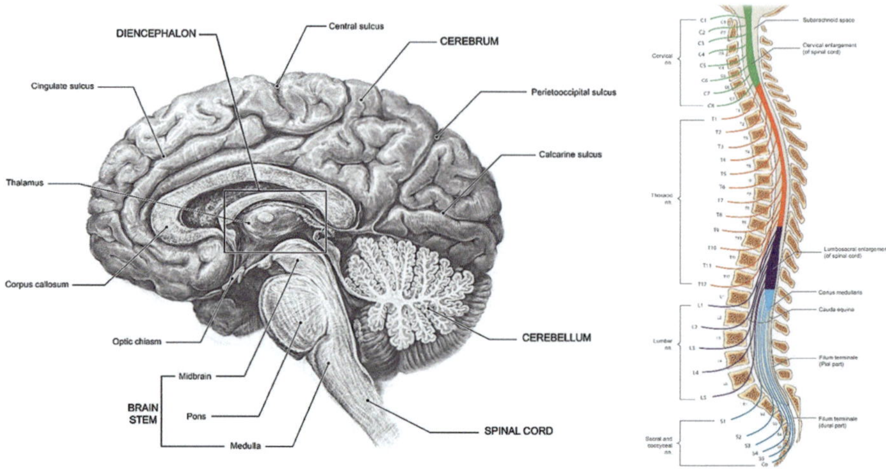

Fig. 6.1 Two parts of the central nervous system. One part (left), the brain located in the skull, is composed by the cerebrum; diencephalon; brain stem consisting of the midbrain, pons, and medulla oblongata; and cerebellum. The other part (right), the spinal cord located in the vertebral foramen, is divided into the cervical, thoracic, lumbar, sacral, and coccygeal portions

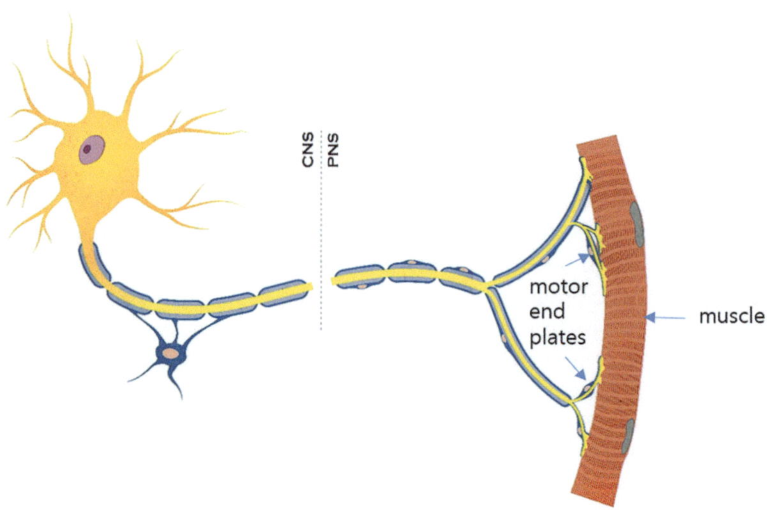

Fig. 6.2 Schematic drawing of the motor nerve cell (motor neuron) and its synapsis with motor end plate. The nerve cells have unique feature characteristic of several long processes for transmission of information generated in their body. *CNS* central nervous system, *PNS* peripheral nervous system muscle

signal is initiated in the first neuron of the brain (cerebrum) and transmitted via the projection fibers of the first neuron to the second neuron located in the spinal cord or the brain stem and then finally it is transferred to the motor end plate in the muscle via the projecting fibers (peripheral motor nerve) of the second neuron. For these functional characteristics, the nervous system is divided into the central nervous system (CNS) and the peripheral nervous system (PNS). The CNS is composed of mainly the regions of neuronal aggregation and the tracts, whereas the PNS includes all kind of the peripheral nerves (Fig. 6.2).

Generally, the brain occupies approximately 2% of the total body weight. Interestingly, it expends approximately 20% of the total oxygen inhaled through the respiratory system and expends roughly 25% of the D-glucose ingested through the GI system. That is the reason why people easily feel hunger when they just study hard without vigorous physical activity. Thus, the brain is one of the organs generating oxygen free radicals most vigorously so that some apparatus for its protection from oxidative damages should be inevitable. The role of antioxidants, therefore, can never be neglected in the brain. Among the many antioxidants functioning in the human body, vitamin C must have pivotal roles because it is water-soluble and its radicals, which are by-products of it, are the least toxic. Interestingly, the vitamin C concentration in neurons is exceptionally high (10 mM) compared to its plasma level (about 50 μM). The exact mechanism involved in the formation of such a great concentration gradient (200:1) between the levels of vitamin C in neurons and plasma has not been fully elucidated yet. There are two routes of entry for vitamin C into the brain: one is through the blood-brain barrier and the other is through the ependymal cell layer in the ventricles of the brain and the central canal of the spinal cord. About 98% of vitamin C in the plasma is present as ascorbate, a reduced form, of which its uptake is usually mediated only by SVCT2. In addition, the endothelial cells of the brain capillaries do not have SVCT2 on their surface but have the glucose transporter (GLUT) so that the reduced form of vitamin C cannot get into the neurons directly from the peripheral blood through the blood-brain barrier. Only the oxidized form, dehydroascorbate (DHA), which comprises just about 2% of the total vitamin C present in the plasma, can be taken up by neurons through GLUT, which is usually expressed on the neuronal surface. Even though there is one more entry route for vitamin C through the CSF pathway in which the reduced form of vitamin C can be taken up by both ependymal cells and neurons which express SVCT2 on their surface, it is actually impossible to explain such a huge concentration gap of vitamin C between neurons and the plasma (Fig. 6.3).

One more thing to consider to reach an answer to such an enigmatic phenomenon is that vitamin C can be recycled from the reduced form to the oxidized form and vice versa. However, nobody knows how many times the recycling can be repeated in the neuronal cytoplasm. This recycling theory of vitamin C can also explain the reason why even people who take very little amount of vitamin C every day can maintain very high concentration of vitamin C up to 10 mM in the neuronal cytoplasm, indicating that vitamin C could be accumulated by recycling without degradation after antioxidant action against ROS profusely generated in numerous neurons during life.

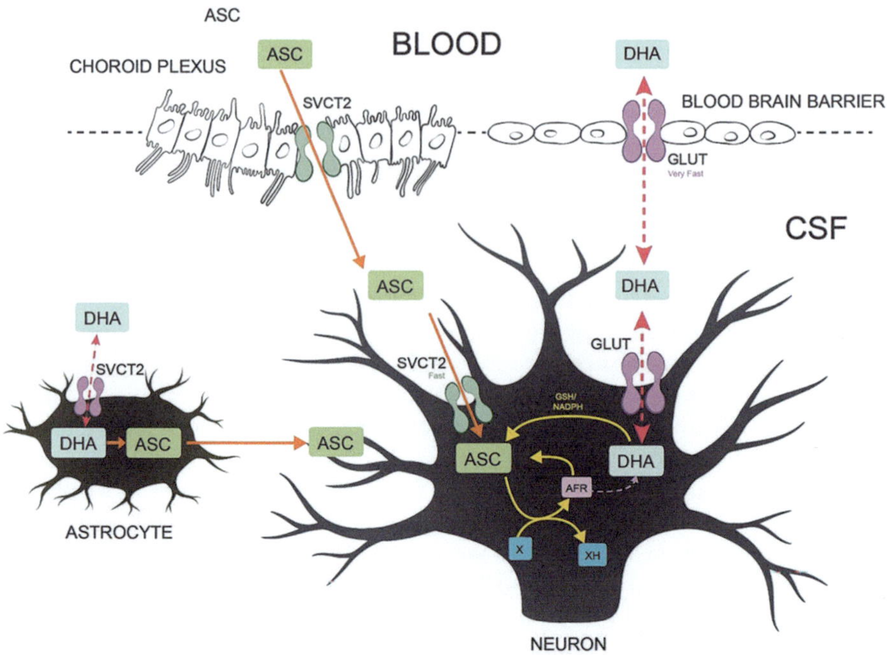

Fig. 6.3 Two routes of entry of vitamin C into the brain. While reduced form of vitamin C, L-ascorbate (ASC), cannot be permeable into the blood-brain barrier (BBB) because the endothelial cells of capillaries do not have SVCT2, oxidized form of vitamin C, dehydroascorbate (DHA), can be permeable into BBB because the endothelial cells of capillaries have the glucose transporters (GLUT). Ependymal cells present on the cerebral ventricles, where CSF has relatively high concentration of vitamin C (200–400 μM) and has SVCT2 on their surface so that ASC in the CSF can be provided to the neurons, which have also SVCT2 on their surface. *AFR* ascorbyl free radical, *CSF* cerebrospinal fluid

Blood-Brain Barrier

Generally, every tissue in the body is supplied with nutrients and oxygen through a microcirculation system mainly composed of capillaries, which permit relatively free permeation of substances in the blood into the surrounding cells or tissues. There are four types of capillaries: type 1 is the continuous capillaries which are found in all kinds of muscle and connective tissues, exocrine glands, and nervous tissues; type 2 is the fenestrated capillaries with a diaphragm, which are found in the kidneys, endocrine organs, and intestines; type 3 is the fenestrated capillaries without a diaphragm, which are found in the renal glomerulus; and type 4 is the discontinuous sinusoidal capillaries mainly distributed in the liver, bone marrow, and spleen. Capillaries consist of a single layer of endothelial cells and pericytes with long processes partly surrounding the endothelial cells. The capillaries distributed over the brain tissue are type 1 and have especially tight junctions between endothelial cells and pericytes in their walls, which usually restrict the free transmission of serum substances and allow permeation of only somc selected materials. Therefore, the permeability of some drugs in the brain capillaries should be considered before their administration. This phenomenon is called the blood-brain barrier, originally suggested by the observation that some drugs given intravenously are not transmitted into the brain tissue, while they easily reach the tissues in areas other than brain. Although the tight junctions play a key role in the function of the blood-brain barrier, it has been known that the footlike long processes of the astrocytes, which form a continuous layer that involves the blood vessels contributing to the blood-brain barrier, also play another role in keeping the barrier function.

Choroid Plexus and Ependymal Cell Layer

The choroid plexus produces the cerebrospinal fluid (CSF), which protects the brain parenchyma from external physical attacks by forming a buffering zone around the brain. Histologically, the CSF is produced in the choroid plexus located in the roofs of the third and fourth ventricles and along the choroid fissure of the lateral ventricles, where the ependymal cell layer comes into contact with the highly vascularized meninges (pia mater), developing into secretory cells and forming the choroid plexus. The CSF produced in the choroid plexus circulates between the ventricles and is distributed in the subarachnoidal space around the brain. One of the apparatuses to protect the brain is the meninx, which is composed of the outermost layer; the dura mater, which is a very strong connective tissue layer; the middle layer; the arachnoid; the innermost layer; and the pia mater, which actually covers intimately the brain surfaces. The subarachnoidal space is formed between the arachnoid membrane and the pia mater (Fig. 6.4).

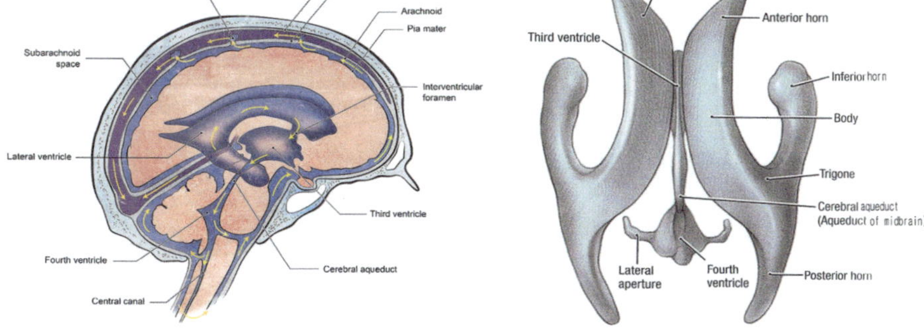

Fig. 6.4 Flow of the cerebrospinal fluid (CSF) in the brain. CSF is produced from the ependymal cells of the choroid plexus located in the roofs of the third and fourth ventricles and along the choroid fissure of the lateral ventricles, where the ependymal cell layer comes in contact with the highly vascularized meninges (pia maters), developing into secretory cells and forming the choroid plexus. CSF produced in the choroid plexus circulates between the ventricles and is distributed in the subarachnoidal space (the space between arachnoid and pia mater) around the brain

Ependymal cells line the ventricles of the brain and the central canal of the spinal cord. Their apical surfaces have many microvilli and one or more cilia. Their basal surface is in contact with astrocytic processes forming the astrocyte end-foot layer. Tanycytes are specialized ependymal cells with basal processes extending between the astrocytic processes to form an end foot on the blood vessels.

Vitamin C and the Brain

Interestingly, the organ containing the highest concentration of vitamin C in the body is the brain. The average concentration of vitamin C in the human peripheral blood is approximately 40~60 μM. However, its concentration in the neuronal cytoplasm reaches approximately 10 mM, about 200 times as high as the human serum concentration (Rice and Russo-Menna 1998; Rice 2000). Additionally, the cytoplasmic concentration of vitamin C in astrocytes or microglial cells, which usually consume 20% of the total amount of oxygen neurons consume, is about 1~2 mM suggesting that the concentration of vitamin C in brain cells seems to be closely related to the consumption amount of oxygen because the amount of oxygen free radicals generated is usually directly proportional to the amount of oxygen consumed (Fig. 6.5).

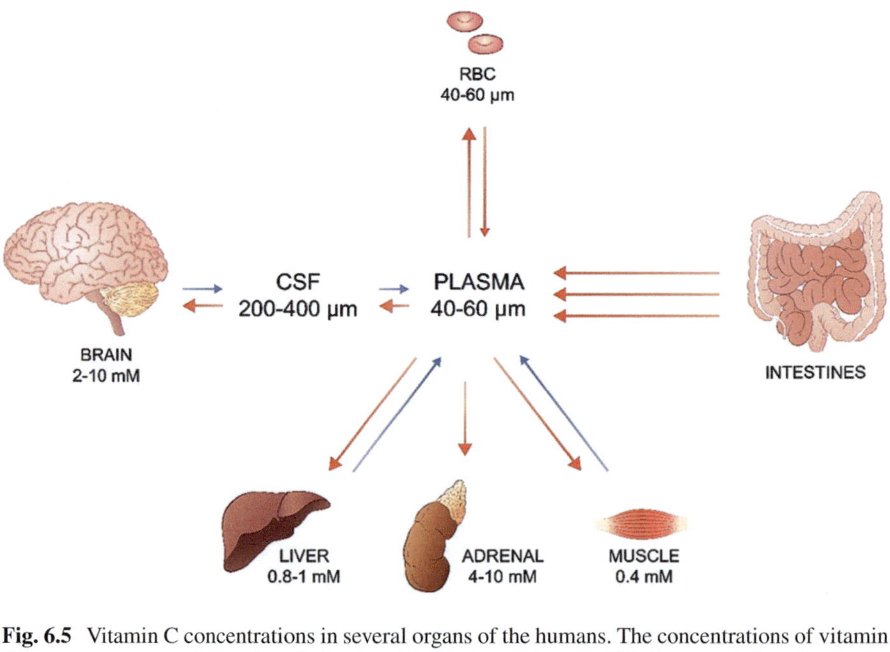

Fig. 6.5 Vitamin C concentrations in several organs of the humans. The concentrations of vitamin C in some organs such as adrenal gland and the brain reach up to 200 times as high as that in the peripheral blood. Such a huge concentration gradient is still under investigation

Neurons are especially sensitive to ascorbate deficiency (Wilson 1997; Hediger 2002). Considering that oxidative damage has a critical role in the pathogenesis of Alzheimer disease or Parkinson disease, the importance of the persistent and regular intake of a sufficient amount of vitamin C can never be overemphasized.

Of interest, the cause of death in scorbutic patients does not seem to be related to the fatal damage of the brain, although scurvy has been known to be associated with paraparesis in humans because vitamin C is avidly retained by the CNS even during ascorbate deficiency (Hornig 1975).

In addition, vitamin C can prevent high glucose-induced apoptosis in human brain pericytes, which is a main cause of cerebral or retinal hemorrhages frequently reported in diabetic patients. High glucose-induced apoptosis is due to the activation of the receptor for advanced glycation end products (RAGE), and its prevention is mediated by the antioxidant function of vitamin C. However, interestingly, its prevention cannot be mimicked by several antioxidants except for vitamin C, suggesting that vitamin C might have a unique function beyond its function as an antioxidant (May et al. 2014).

The concentration of vitamin C in the CSF is about 200–400 μM. As mentioned above, there are two routes for vitamin C entry into the brain: the blood-brain barrier (BBB) and the CSF (Fig. 6.3). In humans, vitamin C gets into the peripheral blood by absorption from ingested food over a small area of the upper jejunum in a SVCT1-mediated manner. About 98% of vitamin C in the peripheral blood is in the reduced form L-ascorbic acid, and only about 2% is in the oxidized form dehydroascorbate (DHA). SVCT can mediate the transport of only reduced vitamin C but not DHA. Endothelial cells in the capillaries consisting of the blood-brain barrier do not express SVCT2 on their luminal surface such that the reduced form of vitamin C in the blood cannot get into the brain through the blood-brain barrier. However, the endothelium of the capillary in the BBB has the glucose transporter (GLUT), which can mediate the transport of DHA but not L-ascorbic acid. On the other hand, SVCT2 is highly expressed on ependymal cells in the choroid plexus (Angelow et al. 2003; Hakvoort et al. 1998; Spector 1977) producing CSF. Therefore, vitamin C in the blood is easily transmitted into the CSF in the ventricles, where vitamin C can also be easily transported into the brain parenchyma by SVCT2, which is also highly expressed on the surface of neurons such that vitamin C in the brain parenchyma easily gets into the neuronal cytoplasm. Despite the second route for vitamin C entry into the brain through the CSF, several questions about the exceptionally huge concentration gradient between the blood and the brain cells still remain to be elucidated. Another thing to consider is how vitamin C is easily recycled in the cytoplasm. L-ascorbate is converted to the L-ascorbyl radical after the antioxidant action, and it is finally converted to DHA after one more antioxidant reaction. However, DHA is not usually degraded to the final metabolites but can be recycled to L-ascorbate in cells through the action of reduced glutathione (GSH) (Fig. 6.6). However, it is not known how many times the recycling of vitamin C can be sustained in cells and even not known how its recycling is regulated.

Now, it has been widely accepted that the vitamin C concentration in the brain is extraordinarily high, although the exact mechanism involved in the formation of

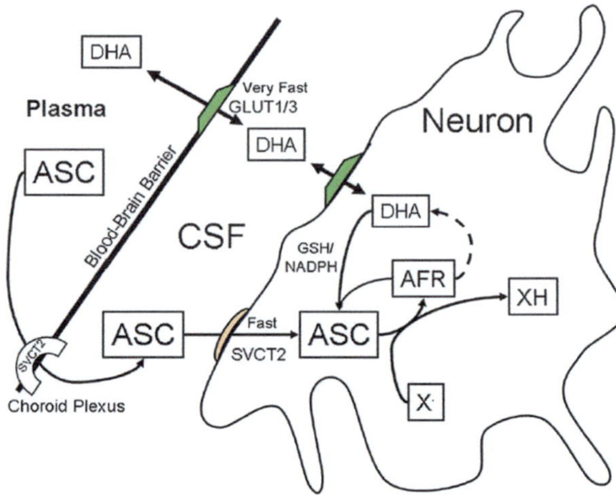

Fig. 6.6 Ascorbate uptake and metabolism in the central nervous system. ASC in the blood can get into the cerebrospinal fluid (CSF) only via the ependymal cells of the choroid plexus and then transported to the neuron through SVCT2. Both the AFR, formed after one-electron donation of ASC, and DHA, transported into the neuron from the blood through GLUT1/GLUT3, are recycled back to ascorbate by glutathione (GSH)/NADPH. *ASC* ascorbate, *AFR* ascorbate free radical, *DHA* dehydroascorbate, *X*∗ oxidizing free radical species. (May 2012)

such a huge concentration gradient has not been fully elucidated. However, interestingly, during the deprivation of vitamin C from the diet for a long time, the vitamin C concentration in the brain decreases very slowly whereas that in the plasma and other tissues decreases rapidly, even resulting in death eventually (Hughes et al. 1971; Hornig 1975). Despite the long discovery history of vitamin C, the exact causes of death in scurvy are still under investigation. However, it is obvious that neurologic problems are minimal even in the dead cases of extreme deficiency of vitamin C (Hirschmann and Raugi 1999), suggesting that at least the neurologic problems caused by the extreme deficiency of vitamin C cannot be the cause of death in scurvy. Normally, because the brain spends a substantial amount of oxygen and thus generates a large amount of oxygen free radicals, it is quite natural that an extremely high concentration of vitamin C is present in the brain to quench the radicals without causing brain damage. Especially in the pathologic state of the brain such as stroke, in which ischemia-reperfusion injury usually occurs, the most drastic and acute oxidant stress takes place. It is generally known well that ischemia exhausts intracellular GSH and ascorbate in the brain and that the following reperfusion generates ROS which further deplete these two antioxidants and extend tissue injury to areas with decreased oxidant defenses (Rice et al. 1995). Well-designed animal studies show that regular supplementation of vitamin C can minimize the infarct size (Ranzan et al. 1993). In similar studies using rats, in which the middle cerebral artery is occluded, only a high dose of dehydroascorbate, but not the reduced form of ascorbate, given intraperitoneally dramatically decreases the infarct volume, mortality, and neurological damages. This indicates that the form of vitamin

C with cerebroprotective effects is the oxidized form dehydroascorbate but not the reduced form ascorbate because only dehydroascorbate is rapidly transported across the blood-brain barrier by glucose transporters (Huang et al. 2001; Agus et al. 1997; Mack et al. 2006), and taken up by glucose transporters, and reduced to ascorbate (May and Asard 2004). In addition, the reason why the reduced form of vitamin C does not have a cerebroprotective effects is because it does not cross the blood-brain barrier in its fully reduced form (Agus et al. 1997; Hosoya et al. 2004). The biochemical background of this phenomenon is that the endothelial cells of the blood-brain barrier lack SVCT2 (Berger et al. 2003; Garcia et al. 2005; Mun et al. 2006).

Another reason to appreciate the importance of the high concentration vitamin C in the brain is cerebral stroke. Brain damage from stroke can be minimized by keeping the vitamin C concentration in the brain high because it has long been known that ischemia releases a substantial amount of vitamin C from brain cells (Hillered et al. 1988) and released vitamin C can prevent excitotoxicity caused by the activation of cell surface and synaptic glutamate receptors through the removal of extracellular glutamate (Grunewald 1993; Rebec and Pierce 1994; Yusa 2001). Surprisingly, vitamin C release and glutamate uptake in neurons and glia have been thought to be closely associated in a mechanism involving heteroexchange between a couple of transporters (Grunewald 1993; Rebec and Pierce 1994; Yusa 2001). However, sometime later, according to several investigations, it was revealed that the actual heteroexchange of two substances is one way-directed. Glutamate can increase ascorbate efflux from cerebral astrocytes, but intracellular ascorbate does not have an effect on the increase of glutamate uptake (Wilson et al. 2000; May et al. 2006).

Furthermore, a high concentration of intracellular vitamin C would be able to quench the ROS generated during glutamate-induced excitotoxicity. Several reports have been published that show vitamin C can prevent glutamate-induced cell damage and death in cultured cerebellar granule cells (Ciani et al. 1996; Atlante et al. 1997). There is another candidate substance causing neuronal excitotoxicity. Activation of the N-methyl-D- aspartate (NMDA) receptor can induce neuronal excitotoxicity, which, fortunately, is known to be prevented by the vitamin C present at a high concentration in the brain cells (MacGregor et al. 1996; Majewska et al. 1990; Majewska and Bell 1990).

Effects of Vitamin C on Neuronal Development

Besides the effect to vigorously quench ROS generated in the brain because of the high consumption of oxygen, it has several effects such as peptide amidation, myelination, and catecholamine synthesis, and the studies on those effects have been sustained for a long time. Just since the beginning of the new millennium, academic evidence on the effects of vitamin C on neuronal development has been forthcoming. In very early 2000, it was shown that vitamin C might have a differentiation inducing potential of embryonic cortical precursor cells into neurons and astrocytes (Lee et al. 2003). In 2006, Qiu et al. reported important experimental results in primary cultures of hippocampal neurons to confirm and extend the findings observed by Lee and colleagues (Qiu et al. 2006). Recently, our team published a paper showing that vitamin C may play a critical role in CNS development (Kim et al. 2015). We utilized vitamin C-deficient Gulo(−/−) mice, which cannot synthesize vitamin C like the humans, because the L-gulono-γ-lactone oxidase-encoding gene was mutated. First, a marked increase in stillbirths and brain damages in neonates from vitamin C-deficient Gulo(−/−) mice was observed. Maybe these findings were caused by an increase in unsolved oxidative stress due to a decrease in vitamin C levels in the brain (Fig. 6.7).

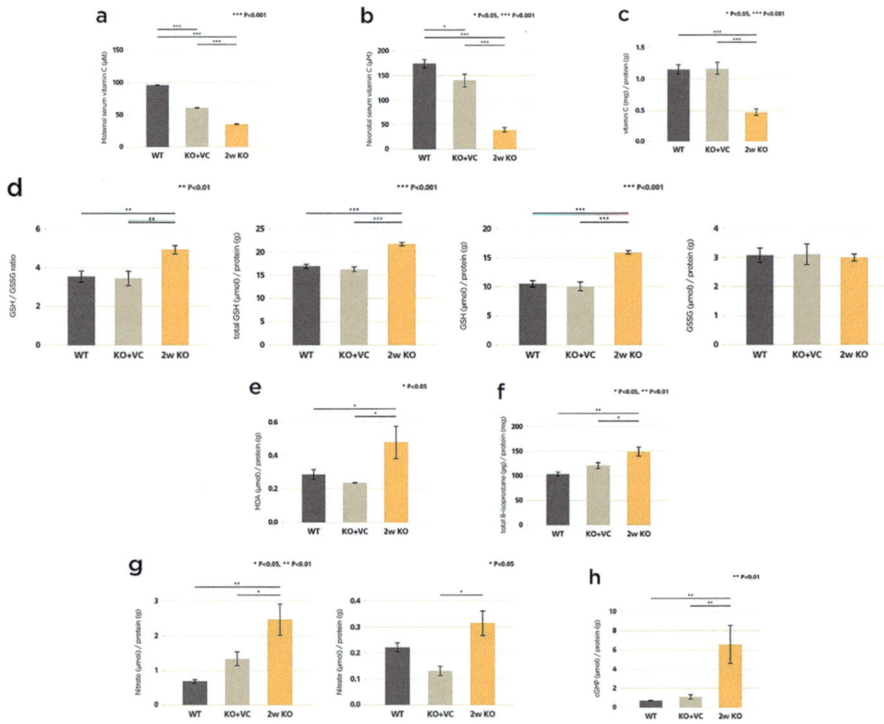

Fig. 6.7 Vitamin C deficiency induces increase in oxidative stress in the neonatal brains of Gulo(−/−) mice. Pregnant Gulo(−/−) mice were depleted of vitamin C for 2 weeks until delivery. (**a**) The lowest blood level of vitamin C in vitamin C-deficient Gulo(−/−) mice (2w KO, n = 8) was observed at the first day of delivery among WT, KO=VC, and 2w KO mice groups. (**b**) Vitamin C concentration in neonatal plasma was also lowest in 2w KO (n = 6) mice at postnatal day (PND) 1. (**c**) Vitamin C concentration in the neonatal brain tissues was lowest in 2w KO mice among WT, KO+ VC, and 2w KO mice (n = 7) at PND 1. (**d**) The level of GSH and GSSG in the neonatal brain tissues was highest in 2w KO mice group among WT, KO=VC, and 2w KO mice groups. The concentrations of (**e**) malonyldialdehyde (MDA) and (**f**) total 8-isoprostane in the neonatal brain tissues were also highest in 2w KO mice group among WT, KO=VC, and 2w KO mice groups. The production of (**g**) nitrate, nitrite, and (**h**) cGMP in the neonatal brain was highest in 2w KO mice group among WT, KO=VC, and 2w KO mice groups. (Kim et al. 2015)

Specifically, we observed a defect in the development of the cerebellum, which indicates an abnormal foliation in the neonatal cerebellum by a decrease in brain-derived neurotrophic factor and glial cell-derived neurotrophic factor expression upon vitamin C deficiency (Fig. 6.8).

Additionally, we also observed defects in motor function upon vitamin C deficiency. We presented morphological evidence for the defects in motor function with atrophic changes of the granule cells and Purkinje cells in the adult cerebellum of Gulo(−/−) mice upon vitamin C deficiency (Fig. 6.9).

Conclusively, vitamin C deficiency during gestation may induce intraparenchymal hemorrhages and severe defects in the development of the cerebellum.

Vitamin C and Neurodegenerative Disorders

It is obvious that the brain consumes 20% of the oxygen inhaled and 25% of the D-glucose ingested resulting in the most powerful generation of oxygen free radicals among all the organs in the body; thus, some mechanism in the brain is indispensable for protection against these toxic radicals. Although many oxygen free radicals are generated in the brain incessantly, average people live without fatal damage to their brain their whole lives, which indicates that there should be some mechanism to protect the brain. Vitamin C is a water-soluble representative antioxidant without any specific or well-known toxicities even at a megadose. The concentration of vitamin C in the neuronal cytoplasm reaches 10 mM, which people or even scientists without in-depth knowledge about vitamin C thought to be a fatally high concentration for any normal cells or tissues. The concentration of vitamin C in glial cells, for which the oxidative metabolism rate is just 1/10 as high as that of neurons (Hediger 2002; Wilson 1997), is just 1–2 mM. Thus, it is strongly suggested that vitamin C should have a special role in the protection of the brain from oxidative damage by oxygen free radicals.

Several neurodegenerative disorders such as Alzheimer disease, Parkinson disease, and Huntington disease are known to be closely related with reactive oxygen species in terms of their pathogenesis (Halliwell 2006).

Dementia

Dementia is a general designation of mental deterioration. Generally, it is classified into two categories, Alzheimer's dementia and vascular dementia. Alzheimer's dementia occupies approximately 70% of all cases, and the remaining cases are of vascular origin. It is well known that oxidative damage is extensively involved in the pathogenesis of Alzheimer's dementia (Montine et al. 2002; Pratico 2002). Furthermore, a lower plasma level of vitamin C (Riviere et al. 1998) as well as a lower level of vitamin C in the CSF (Schipplinget al. 2000) despite an adequate diet

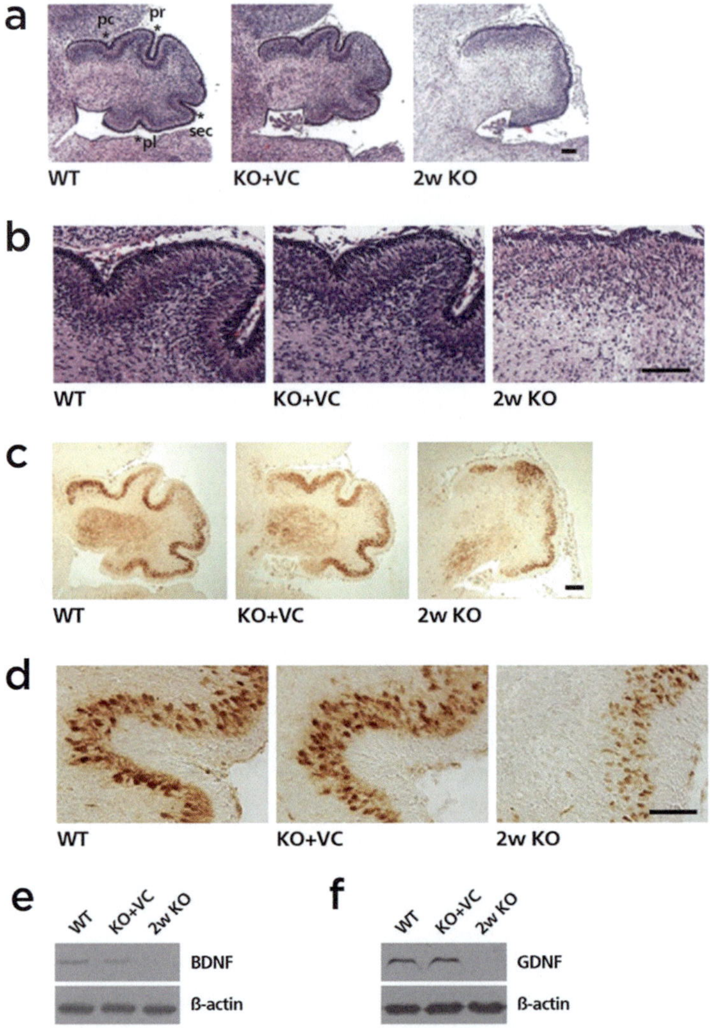

Fig. 6.8 Defect in the cerebellar structure and decreased expression of brain-derived neurotrophic factor (BDNF) and GNDF (glial cell-derived neurotrophic factor) in the neonatal brain of Gulo(−/-) mice upon vitamin C deficiency. At postnatal day (PND) 1, the brains of the neonates from wild-type mice (WT), vitamin C-sufficient Gulo(−/−) mice (KO + VC), and vitamin C-deficient Gulo(−/−) mice depleted of vitamin C for 2 weeks until delivery (2w KO) were examined. (**a**) H&E staining of the neonatal cerebellums. Abnormal foliation and fissure formation were seen in the cerebellum of the neonates only from the 2w KO mice. Principal fissures are marked by asterisks. pc, preculminate; pr, primary; sec, secondary; pl, posterolateral fissure. (**b**) H&E staining of the cerebellum at high magnification. Only in the 2w KO mice, the layer formation of the cerebellar cortex was disturbed. (**c**) Immunohistochemical staining of the neonatal cerebellum with antibody against calbindin-28 k, a marker of Purkinje cell. (**d**) The distribution of Purkinje cells (higher magnification); immunohistochemical staining of the neonatal cerebellum with antibody against calbindin-28 k. (**c**) and (**d**) showed that smaller cell bodies of Purkinje cells and less dendrites in the 2w KO mice were observed than in the WT and KO+VC mice. The expression of (**e**) brain-derived neurotrophic factor and (**f**) glial cell-derived neurotrophic factor was decreased only in the neonatal brain of 2w KO mice. (Kim et al. 2015)

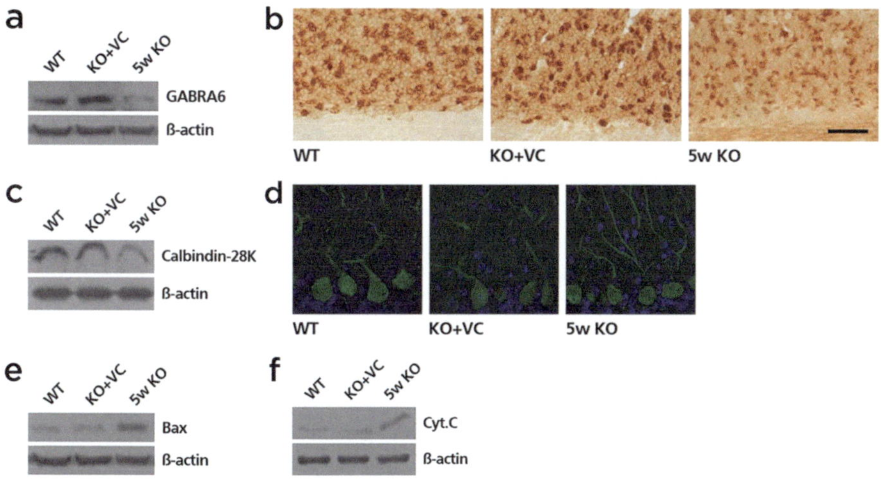

Fig. 6.9 Changes of the granule cells and Purkinje cells in the brain of Gulo(−/−) mice caused by vitamin C deficiency. The supplementation of vitamin C was discontinued for 5 weeks in adult Gulo(−/−) mice, and then the adult brains of the wild-type mice (WT), vitamin C-sufficient Gulo(−/−) mice (KO + VC), and vitamin C-deficient Gulo(−/−) mice (5w KO) were examined. (**a**) Immunoblotting with antibody against GABRA6, a mature granule cell marker, showed that expression of GABRA6 was inhibited by vitamin C deficiency. (**b**) The number and size of mature granule cells were decreased in the cerebellum of 5w KO mice. (**c**) Immunoblotting of calbindin-28 K, a Purkinje cell marker, showed a decrease of calbindin-28 K in the brain of the 5w KO mice. (**d**) Immunofluorescent staining of Purkinje cells with anti-calbindin-28 K antibody (Green) revealed a weak staining in the 5w KO mice. Immunoblotting of (**e**) Bax and (**f**) cytochrome c showed an increased expression of them in the adult brain of the 5w KO mice. (Kim et al. 2015)

has been reported. These papers make me reconsider what the optimal dose of vitamin C for maintenance of normal health including the brain health should be. Even Morris et al. (Morris et al. 1998) and Engelhart et al. (2002) have reported on the positive relationships between supplement use of vitamin C and reduced disease incidence. However, unfortunately, Morris et al. did not know exactly how to administer vitamin C. They used just a single dose of vitamin C daily, as most of the vitamin C researchers have done. As I explained in the introduction, they should have taken vitamin C three times, i.e., every 6 h daily. Fortunately, the beneficial effects of vitamin C could be demonstrated despite a single dose of vitamin C. If they had vitamin C three times a day, they must have had the better results. In addition, there are still several experimental evidences to suggest that vitamin C can have beneficial effects on protecting the brain from the oxidative damage. Rosales-Corral et al. (2003) have demonstrated that the hippocampus area, the main lesion area of Alzheimer disease, could be saved from oxidative damage by orally administering vitamin C. They have also demonstrated that proinflammatory cytokine release induced by injection of β-amyloid peptide, which is well known to be a causative substance of Alzheimer disease, could be reduced. More direct evidence has been proposed by Huang and May (2006) by demonstrating that vitamin C could protect SH-SY5Y neuroblastoma cells from apoptosis and death induced by β-amyloid peptide. In other aspects, there is an additional evidence to support that vitamin C should be beneficial for the management of Alzheimer disease, even if it is not a central one. Vitamin C has been reported to be an effective acetylcholinesterase inhibitor, which is a drug most commonly used for the management of Alzheimer disease (Dhigra et al. 2006).

Parkinson Disease

Parkinson disease is a slowly progressive disease characterized by several motor dysfunctions such as a masklike face, a characteristic tremor of the resting muscles, a slowing of voluntary movements, a festinating gait, a peculiar posture, and a weakness of the muscles. Neurologically, it has been known that a severe decrease of dopaminergic signaling in the central nervous system, among which the substantia nigra containing dopaminergic neurons are known to be a main lesion area, is involved in this disease process. Considering again the relationship between the oxygen consumption proportion (20%) of neurons and the extremely high concentration of vitamin C (10 mM) in the neuronal cytoplasm, oxidative stress seems to have an essential role in this disease process. Unfortunately, there has been a report on a population study showing that vitamin C intake did not protect against the development of Parkinson disease (Zhang et al. 2002). However, the authors of the report did not also have in-depth knowledge of vitamin C such that they just had the populations for the study take a very small single dose of vitamin C daily. My pharmacokinetic study of vitamin C showed a 6-h cycle of the plasma concentration after oral intake (peak at 3 h and basal level at 6 h after oral administration). I had already criticized that the current optimal dose of vitamin C (60~100 mg per day) is not optimal any longer and suggested a new optimal daily dose of vitamin C at 100 times higher the current optimal dose of vitamin C as a new guideline based on scientific background (refer to the introduction part). Therefore, it is no wonder that the positive relationship between vitamin C intake and protection against Parkinson disease development has not been deduced in this population study. Rather, there have been some pharmacological evidences reported that vitamin C can have beneficial effects on the management of Parkinson disease patients. According to the pharmacological study of Nagayama and his colleagues (Nagayama et al. 2004), vitamin C should be very helpful for the management of elderly Parkinson patients with a baseline bioavailability for levodopa because it can improve the bioavailability of levodopa, which can change into dopamine. MPTP (1-methyl-4-phenyl-1,2,36-tetrahydropyridine) is a well-known chemical used for experimental induction of Parkinson disease in the animals. Co-administration of MPTP and vitamin C reduces the toxic effects on the striatum by MPTP by about 20% (Wagner et al. 1986).

Huntington Disease

Huntington disease is also called degenerative chorea, which is characterized by the ceaseless occurrence of a wide variety of rapid, highly complex, jerky movements that appear to be well coordinated but are performed voluntarily. Its pathogenesis is related with the neuronal damages in the striatum maybe caused by the oxidative stress. Rebec and his colleagues reported that mice expressing a gene for this disease exhibited a defect in striatal vitamin C release during periods of behavioral activity (Rebec et al. 2002). Interestingly, not only this defect was recovered by injections of vitamin C but also the behavioral abnormalities of repetitive movements were improved (Rebec et al. 2003).

References

Agus DB, Gambhir SS, Pardridge WM, Speilholz C, Baselga J, Vera JC. Vitamin C crosses the blood-brain barrier in the oxidized form through the glucose transporters. J Clin Invest. 1997;100:2842–8.

Angelow S, Haselbach M, Galla HJ. Functional characterization of the active ascorbic acid transport into cerebrospinal fluid using primary cultured choroid plexus cells. Brain Res. 2003;988:105–13.

Atlante A, Gagliardi S, Minervini GM, Ciotti MT, Marra E, Calissano P. Glutamate neurotoxicity in rat cerebellar granule cells: a major role for xanthine oxidase in oxygen radical formation. J Neurochem. 1997;68:2038–45.

Berger UV, Lu XC, Liu W, Tanf Z, Slusher BS, Hediger MA. Effect of middle cerebral artery occlusion on mRNA expression for the sodium-coupled vitamin C transporter SVCT-2 in rat brain. J Neurochem. 2003;86:896–906.

Ciani E, Groneng L, Voltattorni M, Rolseth V, Contestabile A, Paulsen RE. Inhibition of free radical production or free radical scavenging protects from the excitotoxic cell death mediated by glutamate in cultures of cerebellar granule neurons. Brain Res. 1996;728:1–6.

Dhigra D, Parle M, Kulkarni SK. Comparative brain cholinesterase-inhibiting activity of Glycyrrhiza glabra, Myristica fragrans, ascorbic acid, and metrifonate in mice. J Med Food. 2006;9:281–3.

Engelhart MJ, Geerings MI, Ruitenberg A, van Swieten JC, Hofman A, Witteman JC, Breteler MM. Dietary intake of antioxidants and risk of Alzheimer disease. JAMA. 2002;287:3223–9.

Garcia ML, Salazar K, Millan C, Rodriguez F, Montecinos H, Caprile T, Silva C, Cortes C, Reinicke K, Vera JC, Aguayo LG, Olate J, Molina B, Nualart F. Sodium vitamin C cotransporter SVCT 2 is expressed in hypothalamic glial cells. Glia. 2005;50:32–47.

Grunewald RA. Ascorbic acid in the brain. Brain Res Rev. 1993;18:123–33.

Hakvoort A, Haselbach M, Galla HJ. Active transport properties of porcine choroid plexus cells in culture. Brain Res. 1998;795:247–56.

Halliwell B. Oxidative stress and neurodegeneration: where are we now? J Neurochem. 2006;97(6):1634–58.

Hediger MA. New view at vitamin C. Nat Med. 2002;8:445–6.

Hillered L, Persson L, Bolander HG, Hallstrom A, Ungerstedt U. Increased extracellular levels of ascorbate in the striatum after middle cerebral artery occlusion in the rat monitored by intracerebral microdialysis. Neurosci Lett. 1988;95:286–90.

Hirschmann JV, Raugi GJ. Adult scurvy. J Am Acad Dermatol. 1999;41:895–906.

Hornig D. Distribution of ascorbic acid, metabolites and analogues in man and animals. Ann N Y Acad Sci. 1975;258:103–18.

Hosoya K, Minamizono A, Katayama K, Terasaki T, Tomi M. Vitamin C transport in oxidized form across the rat blood-brain barrier. Invest Ophalmol Vis Sci. 2004;45:1232–9.

Huang J, Agus DB, Winfree CJ, Kiss S, Mack WJ, McTaggart RA, Choudhri TF, Kim LJ, Mocco J, Pinsky DJ, Fox WD, Israel RJ, Boyd TA, Golde DW, Connolly ES Jr. Dehydroascorbic acid, a blood-brain barrier transportable form of vitamin, mediates potent cerebroprotection in experimental stroke. Proc Natl Acad Sci U S A. 2001;98:11720–4.

Huang J, May JM. Ascorbic acid protects SH-SY5Y neuroblastoma cells from apoptosis and death induced by beta-amyloid. Brain Res. 2006;1097:52–8.

Hughes RE, Hurley RJ, Jones PR. The retention of ascorbic acid by guinea-pig tissues. Br J Nutr. 1971;26:433–8.

Kim H, Kim Y, Bae S, Lim SH, Jang M, Choi J, Jeon J, Hwang YI, Kang JS, Lee WJ. Vitamin C deficiency causes severe defects in the development of the neonatal cerebellum and in the motor behaviors of Gulo −/− mice. Antioxid Redox Signal. 2015;23:1270–83.

Lee JY, Chang MY, Park CH, Kim HY, Kim JH, Son H, Lee YS, Lee SH. Ascorbate-induced differentiation of embryonic cortical precursors into neurons and astrocytes. J Neurosci Res. 2003;73:156–65.

MacGregor DG, Higgins MJ, Jones PA, Maxwell WL, Watson MW, Graham DI, Stone TW. Ascorbate attenuates the systemic kainite-induced neurotoxicity in the rat hippocampus. Brain Res. 1996;727:133–44.

Mack WJ, Mocco J, Ducruet AF, Laufer I, King RG, Zhang Y, Guo W, Pinsky DJ, Connolly ES Jr. A cerebroprotective dose of intravenous citrate/sorbitol-stabilized dehydroascorbic acid is correlated with increased cerebral ascorbic acid and inhibited lipid peroxidation after murine reperfused stroke. Neurosurgery. 2006;59:383–8.

Majewska MD, Bell JA, London ED. Regulation of the NMDA receptor by redox phenomena: inhibitory role of ascorbate. Brain Res. 1990;537:328–32.

Majewska MD, Bell JA. Ascorbic acid protects neurons from injury induced by glutamate and NMDA. Neuroreport. 1990;1:194–6.

May JM, Asard H. Ascorbate recycling. In: Asard H, May JM, Smirnoff N, editors. Vitamin C, functions and biochemistry in animals and plants. London: Bios Scientific Publishers; 2004. p. 139–58.

May JM, Jayagopal A, Qu ZC, Parker WH. Ascorbic acid prevents high glucose-induced apoptosis in human brain pericytes. Biochem Biophys Res Commun. 2014;452(1):112–7.

May JM, Li L, Hayslett K, Qu ZC. Ascorbate transport and recycling by SH-SY5Y neuroblastoma cells: response to glutamate toxicity. Neurochem Res. 2006;31:785–94.

May JM. Vitamin C transport and its role in the central nervous system. Subcell Biochem. 2012;56:85–103.

Montine T, Neely M, Quinn J, Beal M, Markesbery W, Roberts L, Morrow J. Lipid peroxidation in aging brain and Alzheimer's disease. Free Radic Biol Med. 2002;33:620–6.

Morris MC, Beckett LA, Scherr PA, Hebert LE, Bennett DA, Field TS, Evans DA. Vitamin E and vitamin C supplement use and risk of incident Alzheimer disease. Alzheimer Dis Assoc Disord. 1998;12:121–6.

Mun GH, Kim MJ, Lee JH, Kim HJ, Chung YH, Chung YB, Kang JS, Hwang YL, Oh SH, Kim JG, Hwang DH, Shin DH, Lee WJ. Immunohistochemical study of the distribution of sodium dependent vitamin C transporters in adult rat brain. J Neurosci Res. 2006;83:919–28.

Nagayama H, Mamamoto M, Ueda M, Nito C, Yamaguchi H, Katayama Y. The effects of ascorbic acid on the pharmacokinetics of levodopa in elderly patients with Parkinson disease. Clin Neuropharmacol. 2004;27:270–3.

Pratico D. Alzheimer's disease and oxygen radicals: new insights. Biochem Pharmacol. 2002;63:563–7.

Qiu L, Li L, Weeber EJ, May JM. Ascorbate transport by primary cultured neurons and its role in neuronal function and protection against excitotoxicity. J Neurosci Res. 2006.

Ranzan A, Theodore D, Haran RP, Chandy MJ. Ascorbic acid and focal cerebral ischemia in a primate model. Acta Neurochir. 1993;123:87–91.
Rebec GV, Pierce RC. A vitamin as neuromodulator: ascorbate release into the extracellular fluid of the brain regulates dopaminergic and glutamatergic transmission. Prog Neurobiol. 1994;43:537–65.
Rebec GV, Barto SJ, Marseilles AM, Collins K. Ascorbate treatment attenuates the Huntington behavioral phenotype in mice. Neuroreport. 2003;14:1263–5.
Rebec GV, Barton SJ, Ennis MD. Dysregulation of ascorbate release in the striatum of behaving mice expressing the Huntington's disease gene. J Neurosci. 2002;22:RC202.
Rice ME, Lee EJ, Choy Y. High levels of ascorbic acid, not glutathione, in the CNS of anoxia-tolerant reptiles contrasted with levels in anorexia-intolerant species. J Neurochem. 1995;64:1790–9.
Rice ME, Russo-Menna I. Differential compartmentalization of brain ascorbate and glutathione between neurons and glia. Neuroscience. 1998;82:1213–23.
Rice ME. Ascorbate regulation and its neuroprotective role in the brain. Trends Neurosci. 2000;23:209–16.
Riviere S, Birlouez-Aragon I, Nourhashemi F, Bellas B. Low plasma vitamin C in Alzheimer patients despite an adequate diet. Int J Geriatr Psychiatry. 1998;13:749–54.
Rosales-Corral S, Tan DX, Reiter RJ, Valdivia-Velazquez M, Martinez-Barboza G, Acosta-Martinez JP, Ortiz GG. Orally administered melatonin reduces oxidative stress and proinflammatory cytokines induced by amyloid beta peptide in rat brain: a comparative, in vivo study versus vitamin C and Eur. J Pineal Res. 2003;35:80–4.
Schippling S, Kontush A, Arit S, Buhmann C, Sturenburg HJ, Mann U, Muller-Thomsen T, Beisiegel U. Increased lipoprotein oxidation in Alzheimer's disease. Free Radic Biol Med. 2000;28:351–60.
Spector R. Vitamin homeostasis in the central nervous system. N Engl J Med. 1977;296:1393–8.
Wagner GC, Carelli RM, Jarvis MF. Ascorbic acid reduces the dopamine depletion induced by methamphetamine and the 1-methyl-4-phenyl pyridinium ion. Neuropharmacology. 1986;25:559–61.
Wilson JX, Peters CE, Sitar SM, Daoust P, Gelb AW. Glutamate stimulates ascorbate transport by astrocytes. Brain Res. 2000;858:61–6.
Wilson JX. Antioxidant defense of the brain: a role for astrocytes. Can J Physiol Pharmacol. 1997;75:1149–63.
Yusa T. Increased extracellular ascorbate release reflects glutamate re-uptake during the early stage of reperfusion after forebrain ischemia in rats. Brain Res. 2001;897:104–13.
Zhang SM, Hernan MA, Chen H, Spiegelman D, Willett WC, Ascherio A. Intakes of vitamin E and C, vitamin supplements, and PD risk. Neurology. 2002;59:1161–9.

Chapter 7
Aging

Although there has been great progress in life sciences such as genetics and molecular biology, many questions about the human life span remain to be unsolved. Actually, it is known that many factors are involved in the aging process; thus, it is difficult to get a clear and simple picture of the aging process.

How long can humans live? Nowadays, the human life span is rapidly increasing. Up to what extent can the human life span be increased? Every animal including humans has a life span. Gerontologists have said that there are three factors that have greatly contributed to prolonging the human life span during the last century: clean drinking water; developed modern medicine, e.g., antibiotics; and improved nutrition.

Actually, a clean drinking water supply system has saved many people from suffering from waterborne infectious diseases. It especially saves young people.

Rapid progress in the techniques and tools of modern medicine such as (1) a vascular stent, which saves many patients who should have died suddenly from a myocardial infarction, and (2) the partial liver graft technique, which also saves many patients with terminal liver cirrhosis or cancer for whom a liver graft has been the only way to save these patients, has been ongoing. In the past, those liver patients all died because it was very hard to get a liver donated because humans only have one liver unlike a kidney. The only source of a whole liver donation was from people with brain death. Therefore, the supply of donated livers could never meet the demands of those liver patients. However, development of the partial liver graft technique has made it possible to donate part of a liver from a healthy son or daughter to the father or mother with a terminal liver disease.

Nutrition has greatly improved with the acceleration of modernization and industrialization. Especially, a sufficient supply of good quality protein has contributed to improving human health despite the increase in metabolic diseases caused by a surplus intake of calories. The importance of the immune system for the maintenance of good health cannot be stressed enough. And almost all the various components of the immune system fighting against foreign invaders are protein-derived substances.

W. J. Lee, *Vitamin C in Human Health and Disease*,
https://doi.org/10.1007/978-94-024-1713-5_7

Definitely, anyone can say the above three factors have greatly contributed to prolonging the life span of humans. However, although modern people meet all three of the abovementioned conditions, eventually, all people die, even if just slowly over a long time without any perception that they are dying. Why do humans still age despite that perfectly ideal conditions for the maintenance of human life are fully satisfied? The so-called physiological aging theory on the normal aging process should be considered, which is different from the pathological aging process, and can achieved by several conditions (mainly the three factors described above) to stay healthy. The physiological aging theory explains the normal aging process but not the pathological aging process (disease processes) occurring to maintain life. However, both processes do not operate independently, but the physiological aging process should be related with the pathological one whenever the principles determining life span are investigated.

Aging Theories

In general, the physiological aging theory can be explained by two categories. One is the gene-based theory, and the other is the non-gene-based theory. Both theories do not work independent of each other but are closely related with each other; thus, I think basically that the non-gene-based theory must operate with the gene-based theory (Fig. 7.1).

The gene-based theory clearly explains why and how each species have their own life span planned in their related genes. For example, mice have a life span of more or less than 3 years, while human beings have a life span more or less than 100 years. However, it is interesting that both murine and human cell lines are usually cultured under the completely same conditions because the exact same mechanisms for the maintenance of life apply to both species. In addition, even very competent life scientists can never tell the difference between human cell lines and murine cell lines using just a light microscope. If so, why do both species have such a great difference in their life spans? Therefore, it is quite natural to speculate that the reason why both species have totally different life spans can be explained based on the differences in genetic background. Actually, there are several gene-based

Fig. 7.1 Two categories of aging theories. One is the gene-based theory, and the other is the non-gene-based theory. The gene-based theory provides answers to why animals in each species have their own definite lifetime, while the non-gene-based theory answers why the animals in a species have different life span. Both theories are not working independently each other, but closely related so that the non-gene-based theory must operate the gene-based theory

Table 7.1 Summary of gene-based aging theories

Programmed longevity	By M. Davidovici et al. (2010)
Hayflick limit theory	By L. Hayflick (1961)
Error and repairs theory	
DNA damage theory	
Redundant DNA theory	
Transcription theory	
Telomere theory	By J. Campisi (2000)

Table 7.2 Summary of none-gene-based aging theories

Cross-linking theory	By Johan Bjorksten (1942)
Hormone theory (endocrine theory)	By van Heemst (2010)
Immunological theory	
Wear and tear theory	By August Weismann (1882)
Inflammation theory	
Oxygen free radical theory	By R. Gerschman et al. (1954), Denham Harman (1956)

aging theories such as programmed longevity, Hayflick limit theory, error and repair theory, DNA damage theory, redundant DNA theory, transcription theory, and telomere theory (Table. 7.1).

However, unfortunately, there have been no clear gene-based theories to elucidate the molecular mechanism involved in the aging process until now. Rather, each gene-based aging theory explains only one part of the aging phenomena. Among them, however, the telomere theory, reported recently in which with aging the telomere is divided dichotomously and finally no more divisions lead to the end of life, is widely accepted. Unfortunately, however, the biochemical mechanism involved in the division of telomeres has not been elucidated yet. The gene-based aging theories can explain why many species have their own different life spans. Nevertheless, it cannot explain why each species has variations in their life spans even within the same species. Actually, some people in the tropical zone have an average life span of 50 years or less, while some people in the high mountain areas usually enjoy a longevity near 100 years. The none-gene-based aging theory can explain why the difference in the average life span between two areas occurs.

In none-gene-based aging theories, there are several theories such as the cross-linking theory, immunological theory, hormone theory (endocrine theory), wear and tear theory, inflammation theory, and oxygen free radical theory (Table. 7.2).

Likewise, all the none-gene-based aging theories except for the oxygen free radical theory just partly explain different aspects of the aging process. However, the oxygen free radical theory is widely accepted as an orthodox of aging theory. This was originally advocated by Dr. Gerschman in 1954 but further developed and matured by Dr. Denham Harman (Gerschman et al. 1954; Harman 1956). This theory tells us that every living subject generates the power to maintain life, for which they intake an energy source and inhale air. The power, which is the basis of life, can

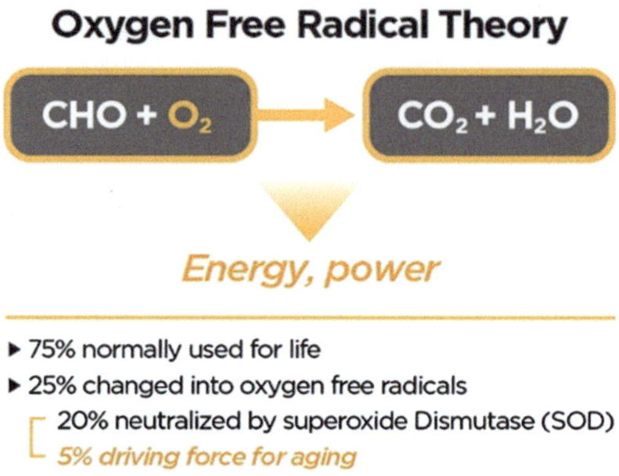

Fig. 7.2 Oxygen free radical theory. Eating and breathing are the most fundamental things for maintenance of life. According to this theory, small portion of oxygen (approximately 25%) inhaled into the lung changed into oxygen free radical, which is known to give oxidative damage to the body. Only 20% of the total oxygen free radical (5% of total oxygen) generated might be a leading cause of aging chronically, while 80% of it (20% of total oxygen) is known to be neutralized endogenously by superoxide dismutase

be generated by the chemical reaction of the energy, source such as carbohydrates and fat, with the oxygen in the inhaled air. During this process, oxygen free radicals are inevitably generated so that every cell in the living body is easily damaged by them, and finally, the living body reaches the end of life. On average, approximately 5% of all oxygen usually used for the maintenance of life has been known to be changed into free radicals (Fig. 7.2).

Considering that aging is "progressive deterioration during whole life of humans that underlies an increasing vulnerability to challenges and a decreasing ability of an organism to survive" (Masoro and Austad 2006), the causes of deterioration lasting during lifelong period should be speculated because the deterioration is usually caused by progressive accumulation of unrepaired somatic damage. Oxidative stress, eventually leading to oxidative damage, must be a representative candidate among many sources causing somatic damage. This opinion is fully supported by numerous scientific evidences over the past few decades that oxidative stress plays a causative role in the pathogenesis of many age-dependent and age-related chronic diseases.

As you can imagine, since the studies supporting the oxygen radical aging theory have had fundamental limitations in using humans themselves, so much research has been done on the effects of oxidative damage on aging in most cases using lower organisms like yeast, worms, flies, and at best mice (Fabrizio et al. 2003; Kenyon 2005; Migliaccio et al. 1999). Interestingly, the results from researches performed thereafter or meanwhile on the effects of oxidative damage in humans are consistent with the results of experiments using lower organisms (Stadtman 1992; Soerensen et al. 2009; Khabour et al. 2009).

Aging and Vitamin C

Many scientists studying aging indicate that a typical phenotype of human aging is frailty, which represents symptoms and signs resulted from prolonged deterioration of each system consisting human body. Specific phenomena related to frailty (Wu et al. 2014) are an age-related muscle loss (called "sarcopenia") (Goodpaster et al. 2001; Delmonico et al. 2007; Morley 2008), an increased adiposity (Hughes et al. 2002; Hubbard et al. 2010; Ouchi et al. 2011), an increased inflammation (Greiwe et al. 2001; Ferrucci et al. 2002, 2005; Mourkioti et al. 2006; Barzilay et al. 2007), age-related hormone decline (Corpas et al. 1993; Herbert 1995; Cappola et al. 2001; Feldman et al. 2002; Bhasin 2008; Voznesensky et al. 2009; Wu et al. 2010), and autonomic nervous system dysfunction (Seals and Esler 2000; Varadhan et al. 2009). Low muscle mass (sarcopenia), weak grip strength, and slow walking speed have been frequently observed in elderly people with vitamin C deficiency, associated with a significant decline in physical function (Kwon et al. 2007a, b; Suzuki et al. 2008). In addition, it is even more important for elderly people to take vitamin C regularly because it has been known that older adults have significantly lower plasma vitamin C levels compared with young adults (Jacob et al. 1988; Blanchard et al. 1990; Fletcher et al. 2003). Since there are so many papers studying effects of oxidative stress or damage on frailty, which consists of several components such as sarcopenia, increased adiposity, hormonal decline including insulin resistance, increased inflammation, and autonomic nervous system dysfunction, representing a human state under abnormalities in multiple systems (Li et al. 1995; Lebovitz et al. 1996; Williams et al. 1998; Lombard et al. 2000; Tyner et al. 2002; Baker et al. 2004, 2011; Chang et al. 2004; Muller et al. 2006), it is quite natural for people to take vitamin C regularly, which is a biologically most ideal antioxidant, for preventing frailty. However, in reality, there have been few papers showing the results that regular intake of vitamin C can delay aging process by preventing oxidative damage in humans. It is actually impossible to perform the intervention study to see antiaging effects of vitamin C using humans because it takes lifelong period. In life science, it is a core fact that several phenomena related with life process should be proven through revealing the real mechanism involved in the process, which can be achieved by the wet-lab experiment. However, the wet-lab experiment has definite limitations, regardless of in vitro or in vivo, so that it is very hard to reflect the real situation. It is even harder to verify real effects of some treatments in case of studies on human health. Therefore, sometimes, some phenomena observed from routine life of huge population can provide some hints so that the scientists make sense of their process and even mechanism involved in it. Following phenomena related with aging process in huge population living in some areas provide some hints to understand the oxygen free radical theory. For example, people who eat a small amount of food usually have a long life span compared to people who eat much. It is quite natural for someone who eats less to consume a small amount of oxygen compared to someone who eats more. Mice deficient in SIR2, a metabolism regulator, have longevity even though the same amount of food was fed to both the SIR2 wild-type and deficient mice, indicating that the mice using less energy live longer than those using more energy. Another interesting example observed in our routine life is that women usually live longer than men by 10% (men vs. women: 78 vs. 85 years). Textbook-wise speaking, men usually inhale more air than women during quiet inspiration. The average volume of inhaled air during quiet inspiration by men is about 650 ml, while it is 500 ml during quiet inspiration by women. Furthermore, a comparison of the number of red blood cells (RBC), which are the blood cells for oxygen transportation, between both sexes provides more convincing evidence that women usually consume less oxygen than men. It is well documented that the number of RBCs in the peripheral blood of men is approximately 5 million/mm^3 while that in women is approximately 4 million/mm^3. Because all men and women living on Earth make a premise to live under the almost same conditions for longevity of life, which include three factors (clean water, well-developed medicine, improved nutrition) described above, it is just the difference in the amount of oxygen free radicals generated during one's life to determine the difference in life span between men and women. Another good example of longevity can be demonstrated in the life of the habitants in high mountain areas. According to the observations of gerontologists, it is well known that there are many centenarians in these areas of high altitudes. It is known that every 1000 m increase in altitude reduces the concentration of oxygen in the air by 5–10%. Therefore, the habitants at high altitudes consume less oxygen although there is a compensation mechanism for the oxygen deficiency, and there are less generation oxygen free radicals resulting in a prolonged life span (Poulain et al. 2004).

Why have there been traditionally no villages with longevity in the tropical zone? It has been known that the average life span of the habitants living in tropical zones is much shorter than even the worldwide mean life span. Generally, living organisms are using more energy in the hot weather than in the cold weather. That is a reason why people are usually easily exhausted during the hot summer season. As such, many Korean people usually have customs to enjoy having special food to provide sufficient energy especially in the summer time. It is well known that the higher the outside temperature is, the higher the basal metabolic rate of living organisms becomes. Therefore, it is not surprising that people living in hot areas use more energy for their increased metabolic rate and, subsequently, more oxygen free radicals are generated resulting in the shortening of their life span over time. Interestingly, the Singaporeans have a higher life span than the worldwide mean life span though they are living in an area on the equatorial line. This paradoxical phenomenon could be possibly explained by the following facts. Singapore is considered a fully developed country. This means that Singaporeans usually have enough food and clean hygiene and have air-conditioning units nationwide so that they are rarely exposed to the hot weather directly during their daily lives. The longevity of the Singaporeans is a very good example to verify that the life span can be increased by artificially modulating the living conditions, especially by air-cooling, although the hot environment makes people use more energy which results in an increase of oxygen free radicals that might result in the shortening of their life spans. There is another good example to show that increased oxygen free radicals can reduce life span.

Traditionally, it is well known that athletes including marathoners have the shortest life span among many occupational groups. They usually use much more oxygen because they live as athletes doing sports their whole lives.

The above examples clearly show how important properly regulating the generation of oxygen free radicals is for longevity. According to oxygen free radical theory, the more oxygen humans consume during their life, the shorter their life span is. There are two ways to cope with health problems caused by oxygen free radicals. One is to reduce the amount of oxygen free radicals generated, and the other is to quench the oxygen free radicals generated with several kinds of well-known antioxidants before they fully damage the body. However, it looks like actually impossible for modern people to reduce the amount of oxygen free radicals because overeating is fatal and extreme stress is commonplace in life for them.

Vitamin C is well known to be the most ideal antioxidant in the biological system. The maintenance of a relatively high level of vitamin C concentration in the peripheral blood of humans can prolong one's life span by reducing the damage caused by oxygen free radicals. Interestingly, Aumailley and his colleagues (2016) reported that the life span of Gulo(−/−) mice, which cannot synthesize vitamin C in their body, increased with the amount of vitamin C in the drinking water from 0.01% to 0.4% vitamin C (w/v) (Aumailley et al. 2016).

Hints from Worldwide Longevity Villages

There are several known areas (Sardinia in Italy, Loma Linda in the USA, Ikaria in Greece, Nicoya Peninsula in Costa Rica, and Okinawa in Japan) in the world where there is a relatively high proportion of old-aged people or centenarians among the general population. Gerontologists say that people living in those areas usually have a couple of common characteristics in their daily life activities. They usually do active physical activities by constantly moving during their daily lives. It suggests that persistent and continuous physical activities with mild intensity, but not extreme physical activities, which cause the generation of excess amount of oxygen free radicals, might effectively delay the aging process. They also have strong social and familial ties, which cannot be made without human relationships based on a warm heart, which can give them peace and love, in other words, a "stress-free life." My research team has reported that a sufficient amount of vitamin C can prevent fatal damage caused by chronic and persistent stress, even save one's life by preventing fatal cardiac damage (Kim et al. 2013). In addition, it is well known that chronic and persistent stress may shorten one's life span. Additionally, it is very hard to find cigarette smokers among them. Thus, many reports have been published on the damages caused by toxic radicals generated during cigarette smoking. Therefore, it is very easy to speculate that cigarette smoking can promote the pathological aging process as well as the physiological one. Considering all these factors, the importance of taking a sufficient amount of vitamin C, as an antioxidant, cannot be overemphasized. These people who live in these longevity regions have optimistic minds. They have healthy dietary habits so that the main contents of their daily diet are composed of vegetables, fruits, and beans, and furthermore, they rarely eat too much.

References

Aumailley L, Warre A, Garand C, Dubois MJ, Paquet ER, Le Couteur DG, Marrette A, Cogger VC, Lebel M. Vitamin C modulates the metabolic and cytokine profiles, alleviates hepatic endoplasmic reticulum stress, and increases the life span of Gulo−/− mice. Aging (Albany NY). 2016;8(3):458–83.

Baker DJ, Jeganathan KB, Cameron JD, et al. BubR1 insufficiency causes early onset of aging-associated phenotypes and infertility in mice. Nat Genet. 2004;36(7):744–9.

Baker DJ, Wijshake T, Tchkonia T, et al. Clearance of p16Ink4a-positive senescent cells delays ageing-associated disorders. Nature. 2011;479(7372):232–6.

Barzilay JI, Blaum C, Moore T, et al. Insulin resistance and inflammation as precursors of frailty: the Cardiovascular Health Study. Arch Intern Med. 2007;167(7):635–41.

Bhasin S. Testicular disorders. In: Korenberg HM, Melmed S, Polonsky KS, Larsen PR, editors. Williams textbook of endocrinology. 11th ed. Philadelphia: Saunders; 2008. p. 645–99.

Blanchard J, Conrad KA, Mead RA, Garry PJ. Vitamin C disposition in young and elderly men. Am J Clin Nutr. 1990;51:837–45.

Cappola AR, Bandeen-Roche K, Wand GS, et al. Association of IGF-I levels with muscle strength and mobility in older women. J Clin Endocrinol Metab. 2001;86(9):4139–46.

Chang S, Multani AS, Cabrera NG, et al. Essential role of limiting telomeres in the pathogenesis of Werner syndrome. Nat Genet. 2004;36(8):877–82.

Corpas E, Harman SM, Blackman MR. Human growth hormone and human aging. Endocr Rev. 1993;14(1):20–39.

Delmonico MJ, Harris TB, Lee JS, et al. Alternative definitions of sarcopenia, lower extremity performance, and functional impairment with aging in older men and women. J Am Geriatr Soc. 2007;55(5):769–74.

Fabrizio P, Liou LL, Moy VN, et al. SOD2 functions downstream of sch9 to extend longevity in yeast. Genetics. 2003;163(1):35–46.

Feldman HA, Longcope C, Derby CA, et al. Age trends in the level of serum testosterone and other hormones in middle-aged men: longitudinal results from Massachusetts Male Aging study. J Clin Endocrinol Metab. 2002;87(2):589–98.

Ferrucci L, Pennix BW, Volpato S, et al. Change in muscle strength explains accelerated decline of physical function in older men and women with high interleukin-6 serum levels. J Am Geriatr Soc. 2002;50(12):1947–54.

Ferrucci L, Corsi A, Lauretani F, et al. The origins of age-related proinflammatory state. Blood. 2005;105(6):2294–9.

Fletcher AE, Breeze E, Shetty PS. Antioxidant vitamins and mortality in older persons: findings from the nutrition add-on study to the Medical Research Council trial of assessment and Management of Older People in the community. Am J Clin Nutr. 2003;78(5):999–1010.

Gerschman R, Gilbert DL, Nye SW, Dwyer P, Fenn WO. Oxygen poisoning and x-irradiation: a mechanism in common. Science. 1954;119:623–6.

Goodpaster BH, Carlson CL, Visser M, et al. Attenuation of skeletal muscle and strength in the elderly: the health ABC study. J Appl Physiol. 2001;90(6):2157–65.

Greiwe JS, Cheng B, Rubbin DC, et al. Resistance exercise decreases skeletal muscle tumor necrosis factor alpha in frail elderly humans. FASEB J. 2001;15(2):475–82.

Harman D. Aging: a theory based on free radical and radiation chemistry. J Gerontol. 1956;11:298–300.

Herbert J. The age of dehydroepiandrosterone. Lancet. 1995;345(8959):1193–4.

Hubbard RE, Lang IA, Llewellyn DJ, Rockwood K. Frailty, body mass index, and abdominal obesity in older people. J Gerontol A Biol Sci Med Sci. 2010;65(4):377–81.

Hughes VA, Frontera WR, Roubenoff R, et al. Longitudinal changes in body composition in older men and women: role of body weight change and physical activity. Am J Clin Nutr. 2002;76(2):473–81.

Jacob RA, Otradovec CL, Russell RM, et al. Vitamin C status and nutrient interactions in a healthy elderly population. Am J Clin Nutr. 1988;48:1436–42.

Kenyon C. The plasticity of aging: insights from long-lived mutants. Cell. 2005;120(4):449–60.

Khabour OF, Abdelhalim ES, Abu-Wardeh A. Association between SOD2 T-9C and MTHFR C677T polymorphism and longevity: a study in a Jordanian population. BMC Geriatr. 2009;9:57–65.

Kim H, Bae S, Kim Y, Cho CH, Kim SJ, Kim YJ, Lee SP, Kim HR, Hwang YI, Kang JS, Lee WJ. Vitamin C prevents stress-induced damage on the heart caused by the death of cardiomyocytes, through down-regulation of the excessive production of catecholamine, TNF-α, and ROS production in Gulo(−/−) Vit C-insufficient mice. Free Radic Biol Med. 2013;65:73–83.

Kwon J, Suzuki T, Yoshida H, et al. Concomitant lower serum albumin and vitamin D levels are associated with decreased objective physical performance among Japanese community-dwelling elderly. Gerontology. 2007a;53:322–8.

Kwon J, Suzuki T, Yoshida H, et al. Association between change in bone mineral density and decline in usual walking speed in elderly community-dwelling Japanese women during 2 years of follow-up. J Am Geriatr Soc. 2007b;55:240–4.

Lebovitz RM, Zhang H, Vogel H, et al. Neurodegeneration, myocardial injury, and perinatal death in mitochondrial superoxide dismutase-deficient mice. Proc Natl Acad Sci U S A. 1996;93(18):9782–7.

Li Y, Huang TT, Carlson EJ, et al. Dilated cardiomyopathy and neonatal lethality in mutant mice lacking manganese superoxide dismutase. Nat Genet. 1995;11(4):376–81.

Lombard DB, Beard C, Johnson B, et al. Mutations in WRN gene in mice accelerate mortality in a p53-null background. Mol Cell Biol. 2000;20(9):3286–91.

Masoro EJ, Austad SN. Handbook of the biology of aging. 6th ed. New York: Academic; 2006.

Migliaccio E, Giorgio M, Mele S, et al. The $p66^{shc}$ adaptor protein controls oxidative stress response and life span in mammals. Nature. 1999;402(6759):309–13.

Morley JE. Sarcopenia: diagnosis and treatment. J Nutr Health Aging. 2008;12(7):452–6.

Mourkioti F, Kratsios P, Luedde T, et al. Targeted ablation of IKK2 improves skeletal muscle strength, maintains mass, and promotes regeneration. J Clin Invest. 2006;116(11):2945–54.

Muller FL, Song W, Liu Y, et al. Absence of CuZn superoxide dismutase leads to elevated oxidative stress and acceleration of age-dependent skeletal muscle atrophy. Free Radic Biol Med. 2006;40(11):1993–2004.

Ouchi N, Parker JL, Lugus JJ, Walsh K. Adipokines in inflammation and metabolic disease. Nat Rev Immunol. 2011;11(2):85–97.

Poulain M, Pes GM, Grasland C, Carru C, Ferrucci L, Baggio G, Franceschi C, Deiana L. Identification of a geographic area characterized by extreme longevity in the Sardinia island: AKEA study. Exp Gerontol. 2004;39:1423–9.

Seals DR, Esler MD. Human aging and the sympathoadrenal system. J Physiol. 2000;528(Pt3):407–17.

Soerensen M, Christensen K, Stevnsner T, Christiansen L. The Mn-superoxide dismutase single nucleotide polymorphism rs1050450 are associated with aging and longevity in the oldest old. Mech Ageing Dev. 2009;130(5):308–14.

Stadtman ER. Protein oxidation and aging. Science. 1992;257(5074):1220–4.

Suzuki T, Kwon J, Kim H, et al. Low serum 25-hydroxy vitamin D levels associated with falls among Japanese community-dwelling elderly. J Bone Miner Res. 2008;23:1309–17.

Tyner SD, Venkatachalam S, Choi J, et al. p53 mutant mice that display early ageing-associated phenotypes. Nature. 2002;415(6867):45–53.

Varadhan R, Chaves PH, Lipsitz LA, et al. Frailty and impaired cardiac autonomic control: new insights from principal components aggregation of traditional heart rate variability indices. J Gerontol A Biol Sci Med Sci. 2009;64(6):682–7.

Voznesensky M, Walsh S, Dauser D, et al. The association between dehydroepiandrosterone and frailty in older men and women. Age Aging. 2009;38(4):401–6.

Williams MD, Van Remmen H, Conrad CC, et al. Increased oxidative damage is correlated to altered mitochondrial function in heterozygous manganese superoxide dismutase knockout mice. J Biol Chem. 1998;273(43):28510–5.

Wu IC, Lin XZ, Liu PF, et al. Low serum testosterone and frailty in older men and women. Maturitas. 2010;67(4):348–52.

Wu IC, Hsiung CA, Hsu CC. Oxidative stress and frailty: a closer look at the origin of a human aging phenotype. In: Preedy VR, editor. Aging-oxidative stress and dietary anti-oxidants. 1st ed. Waltham: Academic; 2014. p. 3–6.

Chapter 8
Stress

Stress is too complicated and vague to define it succinctly. Therefore, actually, when you ask some people what stress is, they usually never have an easy answer. Nevertheless, people cannot only avoid it but also vividly realize its true nature in any moment of their routine life. Sometimes, stress also occurs due to a transition to which people must adapt. By definition, stress means all kinds of forces (external stimuli) that distort the human body and those forces put people into an anxious and threatened state when they confront a situation hard to deal with physically or mentally. Those forces are also called "stressors," the sources inducing stress, while the anxious and threatened state caused by such powers is called "stress responses." It is easy to understand the real meaning of stress, considering that the term, "stress," is originated from the field of architecture. When architects build some structures, they should know what kind of forces will be exerted on the building so that they can prevent the structures from collapsing due to any kind of external force. They even have to calculate the force exerted by the tare weight of a building. Thus, these forces are an essential component to overcome stress. However, for the stress applied to the human body, psychological or mental stress as well as physical stress should be considered. In reality, mental stress is much stronger than physical stress, as modernization in all aspects of society accelerates.

The sources inducing mental stress can be from everything around us. If the routine work repeated every day make you anxious, it is a stressor. If you say "my teacher is a headache," even your teacher can be a stressor. That means that the conditions causing mental stress vary extremely. However, the conditions causing physical stress can be defined relatively clearly. First, all kinds of diseases could be sources of physical stress. Fatigue and overwork are also included in the sources causing physical stress. For example, when people get stung by a bee or catch a cold, their body is under extreme physical stress.

Body responses for stress include psychological ones such as anxiety, restlessness, a sense of fear, sorrow, extreme anger, solicitude, and worry. Sometimes, those responses include physical or physiological ones such as loss of appetite, nausea,

W. J. Lee, *Vitamin C in Human Health and Disease*,
https://doi.org/10.1007/978-94-024-1713-5_8

dyspnea, dry mouth, headache, pain, palpitation, muscle cramps (spasms), insomnia, and loss of concentration.

Body Responses to Stress

There are two motor systems in the body. One is voluntary, which innervates all kinds of skeletal muscles and has a key role in performing all the movements people want to do voluntarily. The other one is the involuntary system which innervates all the smooth muscles located in the walls of internal organs such as the heart, gastrointestinal tract, bronchial trees, and all kinds of vessels. This system is also called the autonomic nervous system because people do not regulate their actions voluntarily. The system mediating the responses for stress is the autonomic nervous system (ANS), a kind of peripheral motor system which consists of the sympathetic and parasympathetic system. Actually, the responses for stress are mainly initiated by the sympathetic system and include an elevated blood pressure, increased heartbeat rate, and elevated blood glucose level (Table 8.1). These stress responses let us know that the body needs more energy when it is in a stressful state. It is understandable that only stable structures (structures built firmly) with strong internal forces, such as buildings, trees, and high-top structures, do not collapse from typhoons, big collisions, etc.; thus, these forces are essential for coping with any kind of stresses. This concept is applied to the human system in exactly the same manner. In the human system, ANS determines the size of the stress to be coped with and then makes the corresponding organs ready to supply a sufficient amount of power. Such roles are performed by the action of the so-called stress-related hormones. The starting point for the responses of the body to overcome stress is the hypothalamus located in the basal part of the brain where it secretes corticotropin-releasing factor (CRF) that stimulates the pituitary gland to secrete adrenocorticotropic hormone (ACTH), which acts on the cortex of the adrenal gland to produce steroid hormones. ACTH also acts on the medulla of the adrenal gland to secrete epinephrine (adrenaline) and norepinephrine (noradrenaline). The core organ with a critical role in coping with stress is the adrenal glands, which are a small paired endocrine organ located on the superomedial aspect of the front of the

Table 8.1 Roles of autonomic nervous system in various functions of body

	Sympathetic	Parasympathetic
Blood pressure	Elevation	Lowering
Heart rate	Increase	Decrease
Sweat in hands	Promotion	Inhibition
Salivation	Inhibition (dry mouth)	Promotion
Air way	Widening	Narrowing
Pupil size	Dilation (mydriasis)	Contraction (miosis)
Bowel movement	Down	Up

corresponding kidney. Histologically, the adrenal gland consists of an outer part called the cortex and an inner part called the medulla. The outer part, the cortex, is divided into three subregions: the most outer region, zona glomerulosa; the middle region, zona fasciculata; and the innermost region, zona reticularis. The zona glomerulosa produces mineralocorticoids, mainly aldosterone which maintains the balance between electrolytes (sodium; Na+ and potassium; K+) and water, thereby keeping the blood pressure within the normal limit. The zona fasciculata and zona reticularis produce glucocorticoids such as cortisone, cortisol, and corticosterone, which regulate carbohydrate, protein, and fat metabolism. Especially under stress, the glucocorticoid hormones usually have a critical role in elevating the blood glucose level. These two zones are also known to secrete androgens (primarily dehydroepiandrosterone) and estrogens in small amounts. The medulla is known to be a part of the peripheral nervous system because it embryologically originates from the neural crest. The medulla produces epinephrine or norepinephrine which is well-known as a stress hormone with inotropic effects on the heart such as elevating the blood pressure and increasing the heart rate (Fig. 8.1).

As I mentioned above, force (energy) is needed to get over stress. When you fall into a stressful state, the pituitary gland (hypophysis) secretes thyroid-stimulating hormone (TSH) more, which stimulates the thyroid gland to produce the thyroid hormone (thyroxine). This hormone has a key role in generating energy within the cytoplasm, thereby elevating the basal metabolic rate. Using the energy generated through this process, stress usually is overcome successfully.

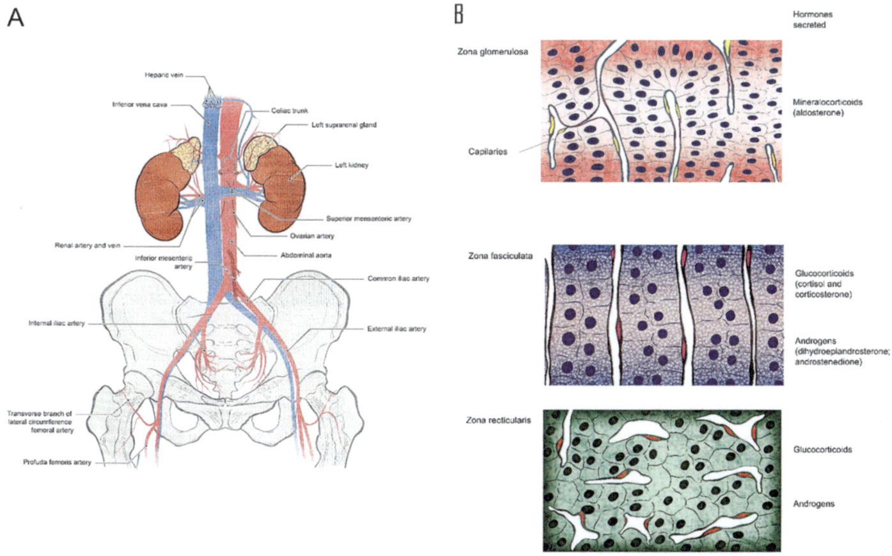

Fig. 8.1 Adrenal glands (**a**) and their topographical organization showing regional patterns of each hormone secretion (**b**)

Role of Vitamin C in the Production of the Stress Hormones

First, it is surprising to know that the adrenal gland is one of the organs containing the highest concentration of vitamin C in the human body. The average concentration of vitamin C in the peripheral blood of humans is about 50 μM. Interestingly, the vitamin C concentration in the adrenal gland is 5–10 mM, which is 100–200 times higher than that in the peripheral blood. Vitamin C is known to have a very important role in the synthesis of stress hormones as a cofactor or co-substrate. In the adrenal cortex, actually, all kinds of steroid hormones are produced, and vitamin C has a role as a co-substrate for the enzyme 7α-monooxygenase which is involved in steroid synthesis (Burri and Jacob 1997). In addition, epinephrine or norepinephrine is produced in the adrenal medulla where vitamin C also has a very critical role as a cofactor in the synthesis of norepinephrine. Specifically, vitamin C is known to be a co-substrate for dopamine-β-monooxygenase, which is an important enzyme acting in the conversion process of the neurotransmitter dopamine to norepinephrine (Carr and Frei 1999). Conclusively, the adrenal gland itself is an organ just for coping with all kinds of stresses through the production of several stress hormones, in which vitamin C has a pivotal role.

Stress and Health

Exposure to stress for a long time has been definitely known to be harmful to maintaining normal health. How can stress have harmful effects on well-being? Major signs shown by the body under a stressful state are an elevated blood pressure and blood glucose level and an increased heart rate, all of which have been known to be closely related with the pathogenesis of several chronic diseases such as atherosclerotic vascular disease and a variety of tumorous diseases, when such a stressful environment has become chronic; however, many other signs and symptoms, which might have harmful effects on health, can be caused by exposure to prolonged stress. Unfortunately, extreme stress looks like a fate destined for modern people in their routine lives because they cannot survive without modern conveniences such as computers and smartphones. While these devices usually make the lives of people easier and more efficient and speedy, they also make them stressful and competitive. Consequently, they must manage their lives in a competitive social environment arising from the use of such modern conveniences. It is even worse, nowadays, because the extreme stress of modern people has become routine. Routine stress in the everyday lives of people indicates that modern people keep both their blood pressure and serum glucose level relatively high (borderline hypertension and diabetes mellitus) over their lifetimes. As I mentioned in the cardiovascular system chapter, prolonged hypertension and a high blood glucose level eventually lead to atherosclerotic vascular disease. Therefore, it can be concluded that the primary bad effect of stress on good health is to induce mild hypertension and diabetes mellitus chronically, which are well-known causes of atherosclerosis. Second, routine stress induces chronic overuse of the energy in the body, leading to exhaustion of energy, followed by functional failure of the body's defense system. A good example of this phenomenon is when people finish a hard, physical task; frequently, they ache all over their bodies or catch a cold. Their immune system is temporarily suppressed by the energy exhaustion that occurred while performing their task. Moreover, the generation of a substantial amount of oxygen free radicals is inevitable in the energy overuse process, which is a key phenomenon that takes place when fighting stress, and those radicals make the first two bad effects (high blood pressure and glucose level) of stress even worse. Furthermore, in the lives of modern people, one more thing that is related to stress and unavoidable is overeating, the primary cause of obesity, that is followed by the generation of excessive oxygen free radicals, which are well known to cause endothelial damage to many arteries, eventually leading to obstructive vascular diseases. If the brain is involved in this case, brain cells are damaged, leading to a potential increased risk of stroke, senile dementia, or Parkinson's disease, and in the case of the heart, a myocardial infarction could be caused by the excessive oxygen free radicals leading to sudden death. Actually, unquenched oxygen free radicals could induce genetic damage, unless repaired properly, leading to cancer formation. Esrefoglu and colleagues reported that neutralization of oxidative stress by antioxidants such as vitamin C and beta-carotene could reduce stress-induced oxidative organ damage in rats. The stresses they adopted were starvation, immobilization, and cold, which gave the rats oxidative stress and caused organ damages, which could be reduced by administration of vitamin C and beta-carotene both inhibiting lipid oxidation and supporting the cellular antioxidant defense system (Esrefoglu et al. 2016).

Utilization of Stress

People use to say that a small amount of stress can be helpful for their health. What does this mean? Actually, we know that a very low level of stress usually refreshes people. The harmful effects of stress on health are primarily when mild hypertension and diabetic status are chronically induced. However, a small amount of stress can elevate the blood pressure and blood glucose level not much beyond the upper limit of their normal ranges. For example, clinically, a normal systolic blood pressure is lower than 130 mmHg. If a low level of stress can maintain the systolic blood pressure around 130 mmHg, people's lives will be more active than those who keep their blood pressure at 80–100 mmHg in their routine lives. This principle can be applied to the blood glucose level as well so that, obviously, people maintaining the upper limit of the normal blood glucose range must live more active lives than those maintaining the lower limit.

Vitamin C and Disease and Death

Many reports have shown that the level of vitamin C concentration in the peripheral blood of patients is generally lower than that of healthy people. Cancer must be a representative disease among the many chronic diseases with a lower vitamin C concentration. In 2009, it was reported that the mean serum vitamin C level was significantly lower in cancer patients than in healthy subjects (Mahdavi et al. 2009). In addition, our unpublished data from liver cancer patients tell us that cancer patients have a much lower level of vitamin C (17.1 ± 12.9 μM, $n = 95$) in their peripheral blood compared to healthy people (42.9 ± 12.4 μM, $n = 44$). And diabetic patients also show the same phenomenon. Another representative disease among chronic infectious diseases is tuberculosis, which is well known to be closely related with vitamin C deficiency (Miric et al. 2013). These reports tell us that the disease itself puts the patient under a kind of physical stress and that condition requires a greater consumption of vitamin C than usual to prevent the damage caused by the stress although the underlying mechanism is not known. Once in a while, extreme stress causes sudden death, for which the exact mechanism has been not elucidated yet. Early in 2000, some scientists reported that the cause of sudden death after an extreme stress might be due to adrenal insufficiency followed by heart failure, using SVCT(−/−) mice, which showed severely reduced tissue levels of vitamin C, impaired dopamine β-hydroxylase activity in the heart, and a prominent catecholamine deficiency in the adrenal gland (norepinephrine was decreased by 50%; epinephrine was decreased by 81%; catecholamine storage vesicles were lost from adrenal chromaffin cells). The mice did not survive 1 month after birth because no cells or organs could use vitamin C for their survival (Bornstein et al. 2003). At that time, this assertion was striking but was not widely accepted because a real-life situation was not reflected by the experiment. In addition, they did not reveal the direct causes of death exactly in vitamin C deficiency, although they demonstrated many symptoms and signs caused by vitamin C deficiency. Our team had performed an interesting in vivo experiment to show that the stress condition could reduce the plasma vitamin C level, in which ultrasonic sound as a stressor was applied to gulo(−/−) mice, which cannot produce vitamin C in their body, and vitamin C or lemon was given to the experimental animals. The results showed that a reduction of plasma vitamin C was observed only in animals to which stress was applied, and the reduced state was recovered by administration of vitamin C or lemon (Fig. 8.2). Kori and his colleagues also reported that restrained stress elevated serum cortisol and reduced serum antioxidant vitamins such as vitamin C and E in experiment using male albino rats and their serum levels were reversed by some drug with an antioxidant property (Kori et al. 2016).

Furthermore, recently, my research team has reported on the scientific reason why and how stress conditions can induce sudden death using gulo(−/−) mice, which do not produce vitamin C due to the lack of a gene encoding the enzyme

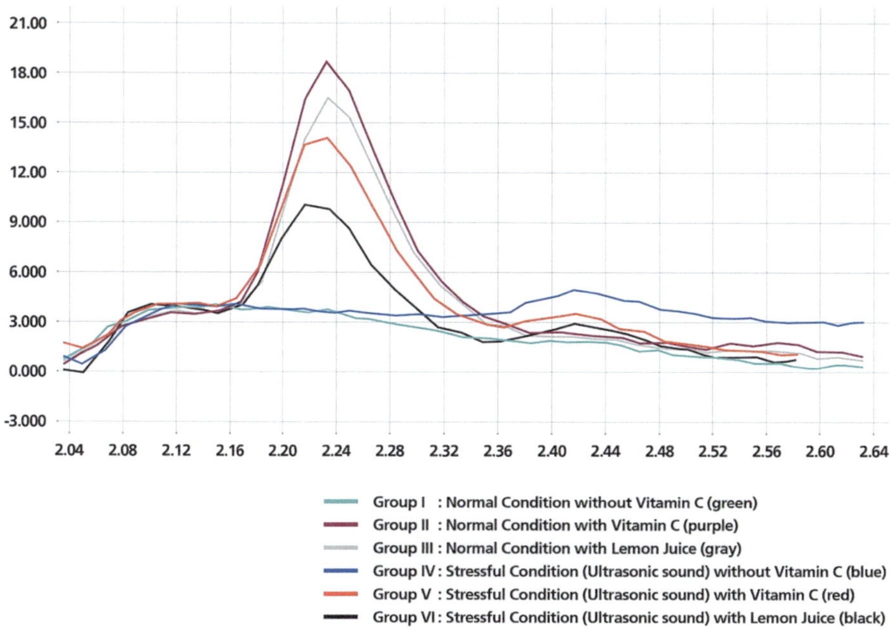

Fig. 8.2 Plasma levels of vitamin C under stress (ultrasonic sound) condition in Gulo(−/−) mice. The plasma levels of vitamin C were increased by oral intake of vitamin C-containing food (vitamin C or lemon juice). However, they were decreased by ultrasonic sound stress (from purple to red or from gray to black) (Our unpublished data)

oxidase for L-gulono-γ-lactone which humans do not have. Two kinds of stress conditions, cold water and restraint, were applied to the mice alternatively to prevent adjustment to the stress from occurring when a single stress condition was applied repeatedly. While the wild mice and gulo(−/−) mice with vitamin C supplementation did not die from the stress, the gulo(−/−) mice without vitamin C supplementation died from the alternative application of the two kinds of stress (Figs. 8.3 and 8.4). The autopsy of the dead mice shows that the cause of death is heart damage (apoptotic death of cardiomyocytes) caused by an excess amount of TNF-α and IL-32 secreted autologously by the cardiomyocytes and p38MAPK is involved in their death pathway (Fig. 8.5). Ironically, the level of stress hormones in the dead mice (the gulo(−/−) mice) is not low under the stress condition but within normal limits, unlike Bornstein's report, which insisted that SVCT(−/−) mice might have died from a lack of stress hormones (Fig. 8.3b) (Kim et al. 2013). A research group by Singal reported that vitamin C could protect cardiomyopathy induced by doxorubicin through serial studies. They found that vitamin C could prevent cardiomyocytes via modulation of nitric oxide synthases from nitrosative stress induced by doxorubicin and by the suppression of the increased TNF-α secretion as well as by

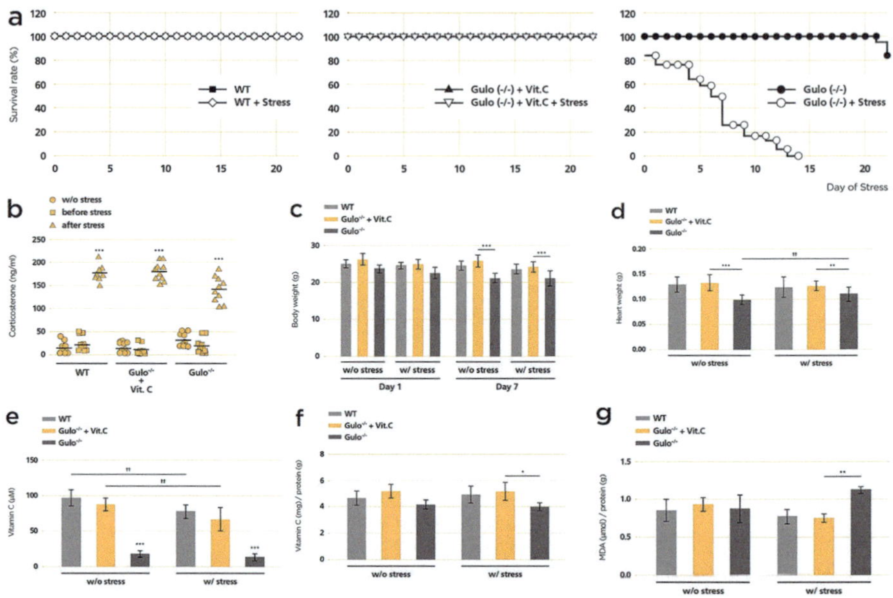

Fig. 8.3 Preventive effect of vitamin C on stress-induced death of Gulo(−/−) mice. (**a**) Gulo(−/−) mice were maintained with or without supplementation of vitamin C for 3 weeks, and then the mortality of WT mice (n = 19), Gulo(−/−) + Vit.C (n = 19), and Gulo(−/−) (n = 22) mice monitored after exposure to stress (1 h/day) showed that there was a high mortality rate only in Gulo(−/−) mice under stress. (**b**) At 5 days of stress exposure, the plasma levels of corticosterone in WT (n = 8), Gulo(−/−) +Vit.C (n = 10), and Gulo(−/−) (n = 10) mice measured by enzyme immunoassay that production of corticosterone was commonly increased by stress in all three groups. (**c**) The body weights of Gulo(−/−) mice maintained with or without supplementation of vitamin C for 3 weeks in WT mice, Gulo(−/−) + Vit. C, and Gulo(−/−) mice were measured at 7 days after stress exposure (n = 9–17). There were no significant changes in the body weights among three groups. (**d**) Heart weights of mice with or without stress exposure were measured (n = 10–12). There were also no significant changes in heart weights among three groups. (**e**, **f**) Vitamin C levels in the (**e**) plasma and (**f**) cardiac tissue measured by an ascorbate assay kit (n = 5/group) showed that the level of vitamin C in the plasma was greatly decreased only in vitamin C-deficient mice, but there was no significant difference in the level of vitamin C in the cardiac tissue among three groups. (**g**) The concentrations of MDA in the cardiac lysates measured by T BARS assay (n = 5–8) showed that the level of MDA was increased only in vitamin C-deficient mice under stress. *MDA* malondialdehyde (Kim et al. 2013)

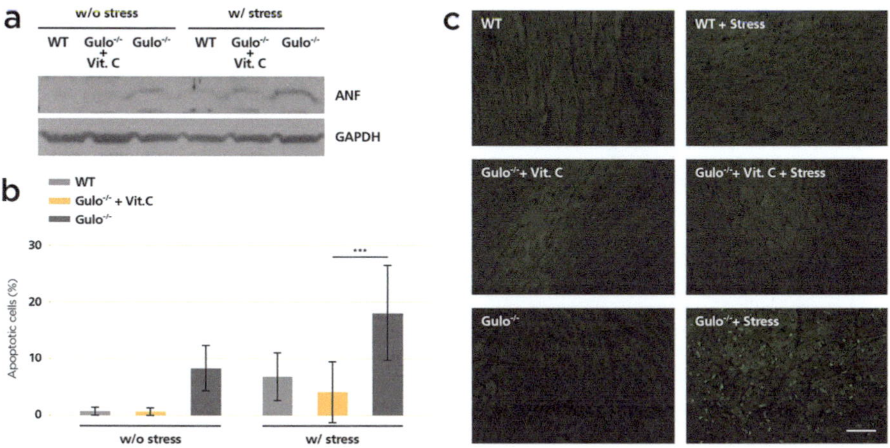

Fig. 8.4 Induction of the extensive apoptosis of cardiomyocytes in Gulo(−/−) mice on vitamin insufficiency and stress. (**a**) ANF expression by immunoblotting in the heart of Gulo(−/−) mice was increased only in vitamin-insufficient mice under stress. (**b**) Apoptotic cells in the heart determined by using an ApopTagPlus peroxidase in situ apoptosis detection kit were markedly increased only in vitamin C-deficient mice under stress. (**c**) The number of apoptotic cells on cardiomyocytes visualized by fluorescent TUNEL (green signals) staining was greatly increased only in vitamin C-deficient mice under stress (Kim et al. 2013)

the reduction of IL-10 secretion in cardiomyocytes induced by doxorubicin (Akolkar et al. 2017). In addition, they reported serially that doxorubicin administration in cardiomyocytes increased the levels of ROS in a time-dependent manner following the activation of p53, p38, and JNK MAPKs and this increase was alleviated by vitamin C adjuvant treatment, suggesting that vitamin C could effectively counteract all doxorubicin-induced changes in cardiomyocytes and might potentially be used as an antioxidant adjuvant therapy to protect against doxorubicin-induced cardiomyopathy (Ludke et.al. 2017).

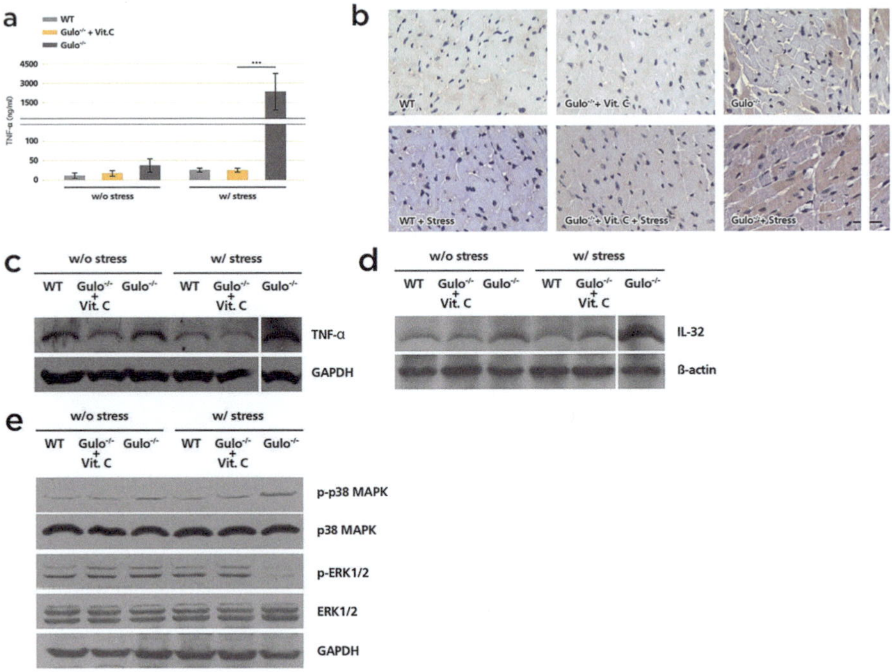

Fig. 8.5 The increased TNF-α and IL-32 production by vitamin C insufficiency and stress in Gulo(−/−) mice through the activation of p38MAPK. The blood and heart were obtained from WT, Gulo(−/−) Vit.C-sufficient (Gulo(−/−)+Vit.C), and Gulo(−/−)Vit.C-insufficient (Gulo(−/−)) mice with or without stress. (**a**) The plasma concentrations of TNF-α measured by enzyme immunoassay showed that TNF-α was greatly increased only in vitamin C-deficient mice under stress among three groups. (**b**) The immunohistochemical staining of TNF-α in the heart showed that there were so many TNF-α-positive cells (brown) observed only in vitamin C-deficient mice under stress. The immunoblotting for expressions of (**c**) TNF-α and (**d**) IL-32 in the heart showed the just same pattern as those observed in (**a**, **b**). (**e**) The immunoblotting for phosphorylated p38MAPK and ERK1/ERK2 in the heart showed that they were more increased in vitamin C-deficient mice under stress than in other groups (Kim et al. 2013)

References

Akolkar G, Bagchi AK, Ayyanppan P, Jassal DS, Singal PK. Doxorubicin-induced nitrosative stress is mitigated by vitamin V via the modulation of nitric oxide synthases. Am J Phys Cell Phys. 2017;312(4):C418–27.

Bornstein SR, Yoshida-Hiroi M, Sotiriou S, Levine M, Hartwig H-G, Nussbaum RL, Eisenhofer G. Impaired adrenal catecholamine system function in mice with deficiency of the ascorbic acid transporter (SVCT2). Faseb J. 2003;17(13):1928–40.

Burri BJ, Jacob RA. Human metabolism and the requirement for vitamin C. In: Packer L, Fuchs J, editors. Vitamin C in health and disease. New York: Marcel Dekker; 1997. p. 341–66.

Carr AC, Frei B. Toward a new recommended dietary allowance for vitamin C based on antioxidant and health effects on humans. Am J Clin Nutr. 1999;69:1086–107.

Esrefoglu M, Akinci A, Taslidere E, Elbe H, Cetin A, Ates B. Ascorbic acid and beta-carotene reduce stress-induced oxidative organ damage in rats. Biochem Histochem. 2016;91(7):455–64.

Kim H, Bae S, Kim Y, Cho CH, Kim SJ, Kim YJ, Lee SP, Hang-Rae K, Hwang YI, Kang JS, Lee WJ. Vitamin C prevent stress-induced damage on the heart caused by the death of cardiomyocytes, through down-regulation of the excessive production of catecholamine, TNF-α, and ROS production in Gulo(−/−) Vit C-insufficient mice. Free Radic Biol Med. 2013;65:573–83.

Kori RS, Aladakatti RH, Desai SD, Das KK. Effect of drug alprazolam on restrained stress induced alteration of serum cortisol and antioxidant vitamins (vitamin C and E) in male albino rats. J Clin Diagn Res. 2016;10(8):AF07–9.

Ludke A, Akolkar G, Ayyappan P, Shrma AK, Singal PK. Time course of changes in oxidative stress and stress-induced proteins in cardiomyocytes exposed to doxorubicin and prevention by vitamin C. PLoS One. 2017;12(7):e0179452.

Mahdavi R, Faramarzi E, Seyedrezazadeh E, Mahammad-Zadeh M, Pourmoghaddam M. Evaluation of oxidative stress, antioxidant status and serum vitamin C in cancer patients. Biol Trace Elem Res. 2009;130(1):1–6.

Miric D, Katanic R, Miric B, Kisic B, Popovic-Katanic N, Nestorovic V. Changes in vitamin C and oxidative stress status during the treatment of tuberculous meningitis. Int J Tuberc Dis. 2013;17(11):1495–500.

Chapter 9
Digestive System

Eating food, nourishment, is essential for maintaining life. During its passage through the gastrointestinal (GI) tract, it is degraded to absorbable forms (e.g., D-glucose, amino acids, and fatty acids) and absorbed into the peripheral blood and then delivered to the cells composing the human body. And finally, it is used in energy production and body maintenance in each cell. The degradation process of food, digestion, and its absorption in the digestive tract are the important essential roles of the GI tract. The digestive system consists of the GI tract and associated glands producing several digestive enzymes and their related substances. Digestion is the process of food degradation, which is divided into mechanical and chemical digestion. Mechanical digestion takes place in the initial stage of digestion for relatively large pieces of food but continues until the moment of absorption. Chemical digestion takes place in the later stage of digestion for relatively smaller pieces of food. The representative actions for mechanical digestion are chewing, peristalsis, and segmentation movement of the GI tract. There are several digestive enzymes for chemical digestion: amylases for carbohydrates (in the oral cavity and pancreas); pepsin (in the stomach), trypsin, and many peptidases (in the pancreas and small intestines) for proteins; and lipase (in the pancreas) for fat.

Gastrointestinal (GI) Tract

The GI tract consists of the esophagus, stomach, small intestines (comprised of three distinct regions, the duodenum, jejunum, and ileum), and large intestines (comprised of four distinct regions, the cecum; ascending, transverse, descending, and sigmoid colons; rectum; and anus) (Fig. 9.1).

Basically the tract is a kind of hollow organ, which has a lumen surrounded by a wall composed of four layers: (1) the mucosa which is the innermost layer actually lining the lumen, (2) submucosa, (3) muscle layer, and (4) serosa which is the outer layer covering the tract. The mucosa has a variety of functions including the

W. J. Lee, *Vitamin C in Human Health and Disease*,
https://doi.org/10.1007/978-94-024-1713-5_9

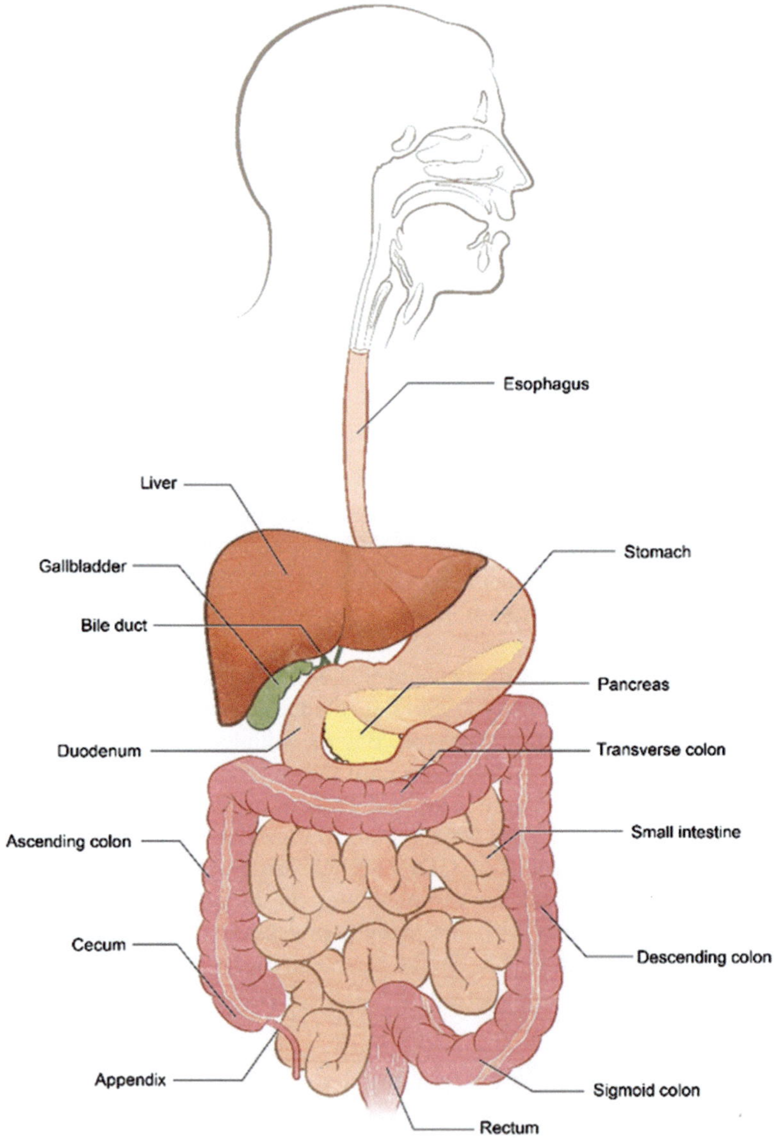

Fig. 9.1 Schematic Drawing of Gastrointestinal (GI) tracts. The GI tracts consist of esophagus, stomach, small intestine (too simplified to be divided into three portions; duodenum, jejunum and ileum), colon (cecum, ascending, transverse, descending and sigmoid colons, rectum)

secretion of fluid and several protein substances and absorption. And actually, there are two muscle sublayers in the muscle layer, the inner circular layer and outer longitudinal layer. Food substances are split and moved forward in the tract by peristalsis and segmentation movement performed by simultaneous contraction of the inner circular and outer longitudinal muscles.

The average length of an esophagus is 25 cm. As an exception to the GI tract, it does not have a serosa as the outermost layer; as such, it is known that esophageal cancer can metastasize to neighboring tissues or organs easily.

The stomach is a sole sac-like organ in the tract and J-shaped with the lesser curvature on its right side and the greater curvature on its left side. The capacity of the stomach is approximately 2.5–3 l. The main function of the stomach is the storage of ingested food. In contrast to the expectations of laymen, the digestive function of the stomach is minimal, and only protein-based food can be digested by pepsin. Amazingly, without any self-damage, the stomach can produce an extremely toxic substance, hydrochloric acid (HCl), which has a couple of functions in digestion: (1) by disinfecting food, it enables people to eat a variety of unsterilized food, and (2) by softening tough food substances, it helps to digest food more easily.

The length of the small intestine is approximately 5–7 m in a dead human body, although some GI physiologists insist that its length is less than 3 m in a living human body. The small intestine carries out the basic and essential functions of the digestive system: chemical digestion and absorption. Actually, digestion is completed in the small intestine in which chemical digestion is really carried out in the duodenum, the first and shortest (approximately 25 cm) part of the small intestine, where all kinds of substances for digestion, such as digestive enzymes and bile juice, are secreted into it from the pancreas and liver. The remaining longer regions of the small intestine, the jejunum and ileum, have primarily absorptive function as well as diminishing the digestive function. There are special structures for effective absorption in the small intestine. Intestinal villi are fingerlike microstructures that develop on intestinal mucosal surface. Because intestinal villi increase the surface area of the mucosa, the absorption efficiency increases greatly compared to the smooth surface of the intestinal mucosa.

The length of the colon is 1.2–1.5 m. The colon is fixed to the posterior wall of the body, whereas the small intestine is relatively free. The colon has three morphological characteristics compared to the small intestine: (1) haustration, (2) three longitudinally oriented bands of strong tissue on its external surface (called taenia coli), and (3) intermittent lipid appendages hanging from its external surface (called appendix epiploicae), which sometimes are difficult for surgeons to differentiate from the vermiform appendix located on the cecum. The primary function of the colon is water absorption, resulting in the formation of solid stools. Another important function of the colon is forming the stool by the action of the normal intestinal microbiota. Actually, the number of microorganisms living in the large intestine is 10–100 times as many as the total number of cells comprising the human body.

Associated Digestive Glands

The digestive system has several associated glands to help in chemical digestion. In the mouth, there are three salivary glands (sublingual, submandibular, and parotid glands), which produce saliva. Saliva is a clear, alkaline, and somewhat viscid substance. The functions of saliva are to moisten and soften food; to keep the mouth moist; to protect against some microbial infections through the actions of lysozymes and lactoferrin, which are mainly secreted from the sublingual and submandibular glands; and, finally, to help in digestion. The digestive function of saliva is only for carbohydrates through the action of amylase (so-called ptyalin), which is primarily produced by the parotid gland. Amylase converts starch into maltose so that people can taste sweetness when they chew starchy foods for a relatively long time.

Once the bolus (ball-like mixture of food and saliva) gets into the stomach, it is readily mixed with the gastric juice produced by some of the cells of the gastric wall through wall-touch stimulation done by the bolus. The gastric juice is comprised of primarily a mucous fluid and hydrochloric acid, which acidifies the bolus, resulting in no more chemical digestion of the carbohydrates by the amylase secreted in the saliva because the optimal pH for salivary amylase is neutral or weakly alkaline. Instead, protein digestion is really initiated by pepsin which is converted from pepsinogen under acidic condition ($pH < 5$). After a 2- to 3-h transit time, the acidified bolus in the stomach is transported into the duodenum, which is the first part of the small intestine, in the order of carbohydrates, proteins, and lipids. Touch stimulation of the duodenum wall by the acidified bolus enables the pancreas to secrete a profuse amount of alkaline pancreatic juice containing mainly sodium bicarbonate ($NaHCO_3^-$), which can usually neutralize the acidity of the gastric bolus. Simultaneously, the pancreas readily produces several digestive enzymes, amylase for carbohydrate digestion, trypsin for protein digestion, and lipase for lipid digestion, for which the optimal pH for these digestive enzymes are all weakly alkaline. Actually, the pancreas has dual functions. While one is an exocrine function, which performs the digestive function for all kinds of foods by producing several digestive enzymes, the other, occupying an extremely small proportion of the pancreas, is the endocrine function. This function is carried out by the production of several hormones in clusters of endocrine epithelial cells known as the islets of Langerhans, such as insulin, glucagon, and somatostatin, among which the representative one is insulin with the important function of lowering blood sugar. Thus, a chronic lack or insufficiency of insulin hormone can cause diabetes mellitus. While the chemical digestion of carbohydrates by amylase gets terminated gradually during its passage through the small intestine, it keeps working for proteins through the action of many peptidases produced in some glands present in the wall of the small intestine as well as by the action of the remaining trypsin. Once the food bolus arrives at the cecum or the ascending colon, the first part of the colon, all functions for chemical and physical digestion are terminated.

The liver is the largest gland in the body. However, the digestive function of the liver is exerted only by the secretion of bile juice, which is indispensable in lipid digestion. Actually, lipid digestion is carried out by lipase produced in the pancreas. In reality, lipids are water-insoluble, and as such, lipase in the watery fluid cannot make contact with lipids, resulting in the failure of lipid digestion. The emulsifying function of the bile juice enables the lipids to contact the aqueous lipase like soap. This is the reason why patients with hepatic diseases, e.g., acute hepatitis, who have a problem producing bile, suffer from fatty stool (steatorrhea). Although bile is produced in the liver, it is transported through an extensive biliary network and gathers together in the gallbladder, where bile is stored and concentrated by the absorption of water and, finally, released into the duodenum through the bile duct in response to the touch stimulation of a fatty diet on the interior wall of the duodenum.

Nitrosamine, a Possible Carcinogen, *H. pylori*, and Vitamin C

The GI tract consists of the esophagus, stomach, small intestine (subdivided into the duodenum, jejunum, and ileum), and large intestine, which actually has no digestive function. The stomach has a unique pocket-like feature so that its primary function is food storage. The stomach has a three-layered muscular tissue band, while the remaining GI tract has a two-layered muscular tissue band (inner circular and outer longitudinal muscular bands). The chemical digestion of proteins by pepsin and mechanical digestion (mixing and slicing) by the three-layered muscular tissue band take place in the stomach. During the mechanical digestion of food, it is known that a possible carcinogen, nitrosamine, might be produced by the chemical reaction of a secondary amine and a nitrous compound (Mirvish and Shubik 1974). Any food may contain these two substances, and as such, it is inevitable for nitrosamine to be produced in the stomach. It is only a matter of how much nitrosamine is produced. Additionally, as people eat more food, more nitrosamine can be produced. These facts can explain why stomach cancer is far more common in men than in women because men tend to eat more food than women. Actually in Korea, the incidence of stomach cancer in men is two to two and half times as high as in women. Fortunately, it is known that vitamin C can prevent the formation of nitrosamine if it is mixed with the food (bolus) that is being digested in the stomach (Weisburger 1977; Mergens et al. 1980; Kalus and Filby 1980; Lintas et al. 1982; Tannenbaum and Wishnok 1987; Tannenbaum 1989; Tannenbaum et al. 1991; Kessler et al. 1992; Yang 1992; Arranz et al. 2007; Erkekoglu and Baydar 2010). Considering that there have been no fundamental treatment modalities for gastric disorders except for anti-acid compounds, which are just helpful for symptom relief, vitamin C could be one of the very good drugs of choice for the treatment of gastric disorders including gastritis and gastric ulcers. This knowledge should be considered when determining how to administer vitamin C. Because vitamin C is a kind of acid, which can cause some chemical damage to some of the lesions found in gastritis or gastric ulcers, it is not recommended to take it in a fasting state, when the stomach is emptied. The best way to take vitamin C is during a meal. Furthermore, it also is known that vitamin C can help in preventing gastric diseases such as gastritis or even gastric cancer caused by *H. pylori.* There have been many studies that have shown the beneficial effects of vitamin C on gastric disorders caused by *H. pylori* (Wang et al. 2000; You et al. 2000; Jarosz et al. 2000; Woodward et al. 2001; Zhang et al. 2002; Feiz and Mobarhan 2002; Simon et al. 2003; Kim et al. 2005). Recent studies on this matter have shown that vitamin C has preventive effects rather than curative ones (Pal et al. 2011).

SVCT1 and Small Intestine

In the GI system, the small intestine has the most important core functions which are digestion and absorption. And it has three subdivisions; the first part, the duodenum, is the shortest part where the pancreatic juice and bile juice are secreted and where the most active digestion occurs. The second part is the jejunum occupying about 40% of the small intestine where active digestion and absorption continue to occur. The third part is the ileum occupying about 60% of the small intestine, and it is connected to the midportion of the cecum, which is the beginning part of the large intestine. Interestingly, vitamin C can be absorbed only through a small part of the small intestine. The pure form of vitamin C (the so-called reducing form of vitamin C) can be absorbed into enterocytes from ingested food through SVCT1, which is mainly distributed over the apical membrane of the small intestine (MacDonald et al. 2002), while SVCT2 is distributed at the basolateral surface, indicating that SVCT2 should have a role in vitamin C transport from the blood into enterocytes (Boyer et al. 2005). Because humans cannot synthesize vitamin C, they inevitably take vitamin C via food intake or vitamin C products directly. Of the total ingested vitamin C, only a very small proportion can be absorbed into the blood vessels through SVCT1 expressed over the apical membrane of the upper jejunum. Because vitamin C can be absorbed over only a limited portion (upper jejunum) of the long GI tract through a specific mechanism (in a sodium-dependent manner), it has generally been known that the absorption rate of vitamin C is very low. Actually, less than 500 mg of vitamin C ingested orally is fully absorbed, while the absorption rate of more than 1000 mg of vitamin C administered orally is no more than 50%. My personal experience suggests that repeated exposure of a relatively high dose of vitamin C might induce enhanced expression of SVCT1 in the upper jejunum. Eventually, it can be said that more than half of the high dose of vitamin C administered orally passes through the small intestine without any intestinal absorption, mixes with the stool that is formed in the colon, and finally is defecated. What is the role of vitamin C passing through the intestine without absorption? Is it just a kind of waste product? The functions of vitamin C in the colon, which is passed through the small intestine without absorption, will be discussed in the next section, "The Colon and Vitamin C."

The Colon and Vitamin C

The main function of the colon is the formation of stool, a solid waste material, through the action of beneficial microorganisms. Therefore, the fundamental difference between the small and large intestines is the absence or presence of the normal intestinal microbiota. Normally, there are no microorganisms in the small intestine, while there are many microorganisms living in the large intestine. Some microbiologists say that humans are living in a sea of microorganisms because it is known that the number of microorganisms living in the large intestine is 10–100 times as many as the total number of the cells comprising the human body (a couple of hundreds of trillion). In addition, the large intestine has neither digestive functions nor an absorption capability for food materials except for water. In view of the clinical aspects, especially the cancer incidence, there should be a very interesting difference between the small and large intestines. Cancer in the small intestines is rare, while cancer in the large intestine is very common. The large intestine has seven regions: cecum, ascending colon, transverse colon, descending colon, sigmoid colon, rectum, and anal canal. The pattern for a cancer attack in each subdivision of the colon has a general trend which shows that if a region is more distally located, the incidence of cancer is higher. Actually, cancers occurring in the anorectal region, the terminal portion of the large intestine, are about half of the total cancer cases in the colon. Considering the differences between the small and large intestines, the higher incidence of colon cancer compared to that for small intestinal cancer might be closely related with the presence of bad microorganisms. Many studies on fat as one of the causative agents of colon cancer have been done such that it has been said that a westernized style diet, which usually is a diet rich in meat containing animal fat, might be the main cause for the increase of colon cancer; furthermore, a possible mechanism has also been reported (Kono 2004; Shao et al. 2015; O'Neill et al. 2016; Penrose et al. 2017). Fat digestion occurs by the bile secreted from the liver via the gallbladder. The bile has a role as an emulsifying agent of fat because it cannot be mixed with water. After fat digestion occurs in the small intestine by the help of the bile juice, the remaining bile is passed through the colon where it is decayed by putrefactive bacteria. Unfortunately, decayed bile has been reported to be a strong carcinogen in the colon. It is known that many microorganisms form the normal intestinal microbiota in the colon, which consists of good fermentative microorganisms and bad putrefactive ones. Based on this, it can generally be speculated that if the proportion of the bad putrefactive bacteria is higher among the normal intestinal microbiota in the colon, there could be a higher incidence of colon cancer although there are some exceptions. My research team has performed an interesting study to verify this speculation. The lactose fermenter assay, which is a microbiological test using culture media containing lactose as an energy source for bacteria instead of glucose, was done for stool samples donated by 20 people, who were divided into 4 groups according to the amount of vitamin C intake: Group 1 consisted of people who took no vitamin C including any supplement products containing vitamin C; Group 2 consisted of people who took 1 g of vitamin C every 6 h

Stool Culture

(lactose fermenter assay)

	Mega-dose of vitamin C			
	>9 gr	6 gr	3 gr	non
lactose fermenter	99	95	80	40
lactose non-fermenter	1	5	20	60

(%)

Fig. 9.2 Lactose Fermenter Assay for human stools after oral intake of mega-dose of vitamin C. The good germs can usually survive by eating lactose or glucose, but the bad germs are impossible to survive with lactose and it is possible to survive only by eating glucose. The lactose fermenter means the proportion of the good germs among total germs in the stool, while the lactose non-fermenter indicates the proportion of the bad germs in the stool. Therefore, this figure shows that as the daily dose of vitamin C increases, the number of good germs in the stool increases (unpublished our data)

with every meal (three times per day, total 3 g); Group 3 consisted of people who took 2 g of vitamin C every 6 h with every meal (three times per day, total 6 g); and Group 4 consisted of people who took 4 g of vitamin C every 6 h with every meal (three times per day, total 12 g). This assay uses the principle that bad putrefactive bacteria cannot live on lactose and require glucose, while the good fermentative ones can live on any kind of energy source including lactose and glucose. Interestingly, Group 4 showed that 99% of the bacterial colonies were alive, indicating that 99% of the colonic microbiota are good fermentative bacteria, and only 1% are bad germs. Group 3 showed that 95% of the bacterial colonies were alive, indicating that 95% of the colonic microbiota are the good fermentative bacteria, and only 5% are bad germs. Group 2 showed that 80% of the bacterial colonies were alive, indicating that 80% of the colonic microbiota are the good fermentative bacteria, and 20% are bad germs. In contrast, Group 1 showed that only 30–40% of the bacterial colonies were alive, indicating that 30–40% of the colonic microbiota are the good fermentative bacteria, while 60–70% are bad germs (Fig. 9.2). These results suggest that vitamin C in the colon, which people usually think is wasted without absorption in the small intestine, might have a role in conversion of bad flora into a good one, such as lactobacilli and nonpathogenic *E. coli* and so on. Recently, a study suggested that a good enteral microbiome has a preventive effect on colorectal cancer (Wang et al. 2018).

Vitamin C and Acute Pancreatitis

The first clue on the effect of vitamin C on acute pancreatitis came from the observation that the plasma levels of vitamin C in patients with acute pancreatitis were much lower than those in the normal control groups (Scott et al. 1993). Bonham et al. even reported that early depletion of vitamin C was closely related with the severity of the acute pancreatitis in a clinical experiment using normal volunteers as well as patients with mild ($n = 62$), and severe ($n = 23$) acute pancreatitis (Bonham et al. 1999). Furthermore, in 2003, Du et al. presented very interesting results to give some clinical guidelines for the management of patients with acute pancreatitis. In this clinical experiment, they showed that a high dose of vitamin C could have therapeutic efficacy on acute pancreatitis, and they even proposed a potential mechanism for these effects, such as promoting an antioxidizing activity in acute pancreatitis patients, blocking lipid peroxidation in the plasma, and improving cellular immune function (Du et al. 2003). Two informative animal studies were done to show that vitamin C could have protective effects against oxidative damage on cerulein-induced pancreatitis (Esrefoglu et al. 2006a, b). In another animal study using 48 Sprague-Dawley rats, Sagiroglu et al. (2008) demonstrated that vitamin C and hypertonic NaCl solution could reduce the amount of necrosis in the pancreas of rats with L-arginine-induced necrotizing pancreatitis (Sagiroglu et al. 2008). Unfortunately, there was a study reporting negative results on the protective effects of vitamin C against oxidative damage in the pancreas. Siriwardena et al. (2007) performed a very meaningful clinical study that was randomized, double-blinded, and placebo-controlled. They gave patients with predicted severe acute pancreatitis 2000 mg of vitamin C in 100 ml of normal saline for the first 2 days and another 1000 mg over the remaining 5 days intravenously, but this study provided no evidence to justify the continued use of vitamin C in severe acute pancreatitis (Siriwardena et al. 2007). Taking a quick look at this report, it seems perfect. Traditionally, however, the starting dose of vitamin C for intravenous injection is at least 10,000 mg. They could have gotten totally different results if they had administered more than 10,000 mg of vitamin C intravenously.

Conclusively, overall, the trends presented in various studies on the protective effects of vitamin C on acute pancreatitis look very promising.

References

Arranz N, Haza AI, Garcia A, Rafter J, Morales P. Protective effect of vitamin C towards N-nitrosamine-induced DNA damage in the single-cell gel electrophoresis (SCGE)/HepG2 assay. Toxicol In Vitro. 2007;21(7):1311–7.

Bonham MJ, Abu-Zidan FM, Simovic MO, Wilkinson A, Winterbourn CC, Windsor JA. Early ascorbic acid depletion is related to the severity of acute pancreatitis. Br J Surg. 1999;86(10):1296–301.

Boyer JC, Campbell CE, Sigurdson WJ, Kuo SM. Polarized localization of vitamin C transporters, SVCT1 and SVCT2, in epithelial cells. Biochem Biophys Res Comm. 2005;334:150–6.

Du WD, Yuan ZR, Sun J, Tang JX, Cheng AQ, Shen DM, Huang CJ, Song XH, Yu XF, Zheng SB. Therapeutic efficacy of high-dose vitamin C on acute pancreatitis and its potential mechanisms. World J Gastroenterol. 2003;9(11):2565–9.

Erkekoglu P, Baydar T. Evaluation of the protective effect of ascorbic acid on nitrite- and nitrosamine-induced cytotoxicity and genotoxicity in human hepatoma line. Toxicol Mech Methods. 2010;20(2):45–52.

Esrefoglu M, Gul M, Ates B, Batcioglu K, Selimoglu MA. Antioxidative effect of melatonin, ascorbic acid and N-acetylcysteine on caerulein-induced pancreatitis and associated liver injury in rats. World J Gastroenterol. 2006a;14(2):259–64.

Esrefoglu M, Gul M, Yilmaz I. Ultrastructural clues for the protective effect of ascorbic acid and N-acetylcysteine against oxidative damage on caerulein-induced pancreatitis. Pancreatology. 2006b;6(5):477–85.

Feiz HR, Mobarhan S. Does vitamin C intake slow the progression of gastric cancer in Helicobacter pyroli-infected populations? Nutr Rev. 2002;60(1):34–6.

Jarosz M, Dzieniszewski J, Dabrowska-Ufniarz E, Wartanowicz M, Ziemlanski S. Tobaco smoking and vitamin C concentration in gastric juice in healthy subjects and patients with Helicobacter pyroli infection. Eur J Cancer Prev. 2000;9(6):423–8.

Kalus WH, Filby WG. Inhibition of nitrosamine formation by ascorbic acid: participation of free radicals in its anaerobic reaction with nitrite. Experientia. 1980;36(2):147–9.

Kessler H, Husenmann B, Wagner W. Potential protective effect of vitamin C on carcinogenesis caused by nitrosamine in drinking water: an experimental study on Wistar rats. Eur J Surg Oncol. 1992;18(3):275–81.

Kim DS, Lee MS, Kim YS, Bae JM, Shin MH, Ahn YO. Effect modification by vitamin C on the relation between gastric cancer and Helicobacter pylori. Eur J Epidemiol. 2005;20(1):67–71.

Kono S. Secular trend of colon cancer incidence and mortality in relation to fat and meat intake in Japan. Eur J Cancer Prev. 2004;13(2):127–32.

Lintas CL, Clark A, Fox J, Tannenbaum SR, Newberne PM. In vivo stability of nitrite and nitrosamine formation in the dog stomach: effect of nitrite and amine concentration and of ascorbic acid. Carcinogenesis. 1982;3(2):161–5.

MacDonald L, Thumser AE, Sharp P. Decreased expression of the vitamin C transporterSVCT1 by ascorbic acid in a human intestinal epithelial line. Br J Nutr. 2002;87:97–100.

Mergens WJ, Chau J, Newmark HL. The influence of ascorbic acid and DL-alpha-tocopherol on the formation of nitrosamines in an in vitro gastrointestinal model system. IARC Sci Publ. 1980;31:259–69.

Mirvish SS, Shubik P. Letter: ascorbic acid and nitrosamine. Nature. 1974;252(5479):179.

O'Neill AM, Burrington CM, Gillaspie EA, Lynch DT, Horsman MJ, Greene MW. High-fat western diet-induced obesity contributes to increased tumor growth in mouse models of human colon cancer. Nutr Res. 2016;36(2):1325–34.

Pal J, Sanal MG, Gopal GJ. Vitamin C as anti-Helicobacter pylori agent: more prophylactic than curative-critical review. Indian J Pharmacol. 2011;43(6):624–7.

Penrose HM, Heller S, Cable C, Nakhoul H, Baddoo M, Flemington E, Crawford SE, Savkovic SD. High-fat diet induced leptin and Wnt expression: RNA-sequencing and pathway analysis of mouse colonic tissue and tumors. Carcinogenesis. 2017;38(3):302–11.

Sagiroglu T, Eren E, Tunca F, Meydan B, Ertekin C. The effects of hypertonic saline solution, ascorbic acid and low-molecular-weight heparin on acute necrotizing pancreatitis in rats. Eurasian J Med. 2008;40(2):53–7.

Scott P, Bruce C, Schofield D, Shiel N, Braganza JM, McCloy RF. Vitamin C status in patients with acute pancreatitis. Br J Surg. 1993;80(6):750–4.

Shao S, Kao TC, Eckhaus J, Bourgeois J, Perera K, Zhu K. The association of percentage energy from fat and colon cancer risk among members of the US military. Eur J Cancer Prev. 2015;24(3):188–94.

Simon JA, Hudes ES, Perez-Perez GI. Relation of serum ascorbic acid to Helicobacter pylori serology in US adults: the Third National Health and Nutrition Examination Survey. J Am Coll Nutr. 2003;22(4):283–9.

Siriwardena AK, Mason JM, Balachandra S, Bagul A, Galloway S, Formela L, Hardman JG, Jamdar S. Randomized, double blind, placebo controlled trial of intravenous antioxidant (N-acetylcysteine, selenium, vitamin C) therapy in sever acute pancreatitis. Gut. 2007;56(10):1439–44.

Tannenbaum SR. Preventive action of vitamin C on nitrosamine formation. Int J Vitam Nutr Res Suppl. 1989;30:109–13.
Tannenbaum SR, Wishnok JS. Inhibition of nitrosamine formation by ascorbic acid. Ann N Y Acad Sci. 1987;498:354–63.
Tannenbaum SR, Wishnok JS, Leaf CD. Inhibition of nitrosamine formation by ascorbic acid. Am J Clin Nutr. 1991;53(1suppl):247S–50S.
Wang X, Willen R, Wadstrom T. Astaxanthin-rich algal meal and vitamin C inhibit Helicobacter pylori infection in BALB/cA mice. Antimicrob Agents Chemother. 2000;44(9):2452–7.
Wang CZ, Huang WH, Zhang CF, Wan JY, Wang Y, Yu C, Williams S, He TC, Du W, Musch MW, Chang EB, Yuan CS. Role of intestinal micribiome in American ginseng-mediated colon cancer prevention in high fat diet-fed AOM/DSS mice [corrected]. Clin Transl Oncol. 2018;20(3):302–12.
Weisburger JH. Vitamin C and prevention of nitrosamine formation. Lancet. 1977;2(8038):607.
Woodward M, Tunstall-Pedoe H, McColl K. Helicobacter pylori infection reduces systemic availability of dietary vitamin C. Eur J Gastroenterol Hepatol. 2001;13(3):233–7.
Yang WX. Exposure level of N-nitrosamine in the gastric juice and its inhibition by vitamin C in high risk areas of esophageal cancer. Zhonghua Zhong Liu Za Zhi. 1992;14(6):407–10.
You WC, Zhang L, Gail MH, Chang YS, Liu WD, Ma LJ, Li YJ, Jin ML, Hu YR, Yang CS, Blaser MJ, Correa P, Blot WJ, Fraumeni JF, Xu GW. Gastric dysplasia and gastric cancer: Helicobacter pylori, serum vitamin C, and other risk factors. J Natl Cancer Inst. 2000;92(19):1607–12.
Zhang ZW, Abdullahi M, Farthing MJ. Effect of physiological concentrations of vitamin C on gastric cancer cells and Helicobacter pyroli. Gut. 2002;50(2):165–9.

Chapter 10
Liver

The liver is the single largest organ in the human body. Its weight varies from 1000 to 3000 g. In fact, it is also the largest exocrine gland, and it produces bile juice. Hence, the liver has wide network for bile passage, composed of the bile canaliculi, bile ductules, and bile ducts. Bile flows in an inside-out manner, whereas the hepatic artery and portal vein flows are the outside-in type. The shape of the liver is a hemi-hexahedron, which results when cutting a hexahedron obliquely. It is located just beneath the thoracic diaphragm, shifted to the right side. While just beneath the liver on the right side, the right kidney is located, and on the left side, the lower surface of the liver covers the stomach and the connection between the stomach and the duodenum. The liver has several important functions. First, it functions as a storage site of essential nutrients such as carbohydrates (as glycogen) and lipids (as triglycerides) so that it can supply the body with energy between meals. That is why the venous blood from the gastrointestinal (GI) tract, which contains all types of nutrients absorbed in the tract, comes into the liver through the hepatic portal vein, except for complex lipids (chylomicron), which are transported mainly by the lymph vessels and distributed throughout the hepatic area through the re-capillarization of the hepatic portal vein, though it is a type of vein. This is an anatomically meaningful exception because by definition all veins become larger in the process of merging at the heart via one of the largest veins, the inferior or superior vena cava, while all arteries become smaller, eventually becoming capillaries. When people starve, the venous blood, without nutrients, when passing through the GI tract, obtains nutrients stored in hepatocytes and delivers them to heart. When people have something to eat, the venous blood passing through the tracts, in this case with numerous nutrients, stores them in the hepatocytes in case of starvation, delivering only part of them to the heart, so that the peripheral blood keeps the level of the nutrients constant to that they can be supplied promptly when needed. The liver also serves as a major storage site for vitamins, especially vitamin A. It produces bile juice, which is stored in the gallbladder, where it is concentrated and secreted into the duodenum upon stimulation with the fat component contained in ingested food. Bile juice is essential for fat digestion in the small intestine because it mixes

W. J. Lee, *Vitamin C in Human Health and Disease*,
https://doi.org/10.1007/978-94-024-1713-5_10

water-insoluble fat with a water-soluble lipase secreted from the pancreas in a process known as emulsification. That is why some patients with the hepatic disorder suffer from the steatorrhea, which means a light stool floats because the stool contains much undigested fat due to a lack of bile juice. Because hepatic bile production is a type of exocrine function, some hepatic disorders in which the hepatic tissues have been damaged usually show symptoms of jaundice, characterized by hyperbilirubinemia and a deposition of bile pigment in the skin and the mucous membranes, resulting in a yellow appearance of the patient. Jaundice is caused by bile contamination of the peripheral blood. Normally, the bile component cannot be present nor be should present in peripheral blood. Furthermore, the liver produces very important plasma proteins, such as albumin; fibrinogen; prothrombin; coagulation factors V, VII, and IX; lipoproteins; and globulins. Albumin is a very important protein for maintaining the osmotic pressure of the blood, which prevents the fluid component of the blood from leaking out of the blood vessels and maintains the blood pressure within its normal limits. This explains why liver cirrhosis patients frequently suffer from ascites, in which the intravascular fluid escapes into the extravascular space, especially into the peritoneal cavity. In addition, they show prolonged bleeding due to the insufficient production of fibrinogen and coagulation factors. The liver also has the function of converting toxic nitrogen compounds such as ammonia into nontoxic compounds, specifically urea. After the metabolism of proteins, typical nitrogen-containing nutrients, the toxic nitrogen compound ammonia is produced and usually converted into the nontoxic nitrogen compound urea in the liver and excreted through the kidneys. In the liver, detoxification and inactivation of various drugs and substances are processed by oxidation, methylation, or conjugation. For example, glucuronyl transferase, an enzyme conjugating glucuronic acid to bilirubin, also causes the conjugation of several other compounds, such as steroids, barbiturates, antihistamines, and anticonvulsants, resulting in their inactivation (Fig. 10.1).

Blood Supply System to the Liver

In the human body, there are two organs which extraordinarily have a dual blood supply system. One is the lungs, to which two arteries, the pulmonary artery and the bronchial artery, supply blood. The other is the liver, which is supplied with blood dually through the original route of the blood supply to the liver, the hepatic artery, and the portal vein, which is one of the large veins and which drains blood from the GI tract. The portal vein can actually supply the liver with oxygen and nutrients, although it is a large vein, even in cases when the hepatic artery is obstructed.

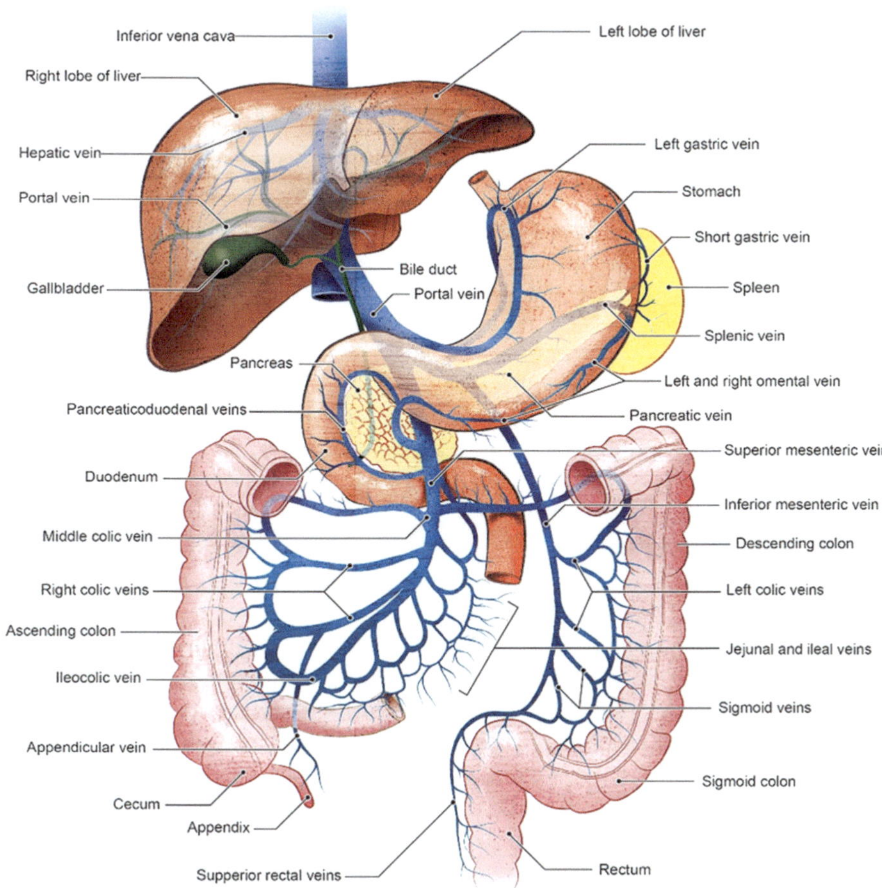

Fig. 10.1 Liver and portal venous system. All blood supplied to the GI tracts is collected into the portal vein and enters the liver, where the portal vein, although it is a kind of the vein, is exceptionally divided into numerous capillaries and distributed all over the liver

Liver and Immune System

The liver is not usually considered to be an organ related to the immune system. Nevertheless, it has definite immune functions. Substantial amounts of immunoglobulins are known to exist in the liver, usually in association with chronic liver diseases (Trevisan et al. 1983; De Brito et al. 1985). An antigen injected into the portal vein does not induce active immune responses, even when the same antigen provokes violent immune responses if injected into other systemic veins. This result indicates the liver has a regulatory function over immune responses. This is the immunological basis for the oral tolerance of various protein compounds, which usually provoke vigorous immune reactions if they are injected into the vessels.

Among the candidate cells taking part in the immune responses of the liver are Kupffer cells, part of the mononuclear phagocyte system (MPS).

Vitamin C and Liver Diseases

My father-in-law suffered from liver cirrhosis, a typical chronic liver disease prevailing in Korea due to the endemic of type B viral hepatitis. In 1988, he suffered a rupture of the esophageal varix, a terminal symptom of liver cirrhosis. Esophageal venous varicosities are usually caused by excessive bypass venous flows of the esophageal vein from tributaries of the portal vein in patients with portal hypertension, of which the cause is increased back pressure caused by instances of intrahepatic venous narrowing or obstruction. However, surprisingly, his esophageal bleeding was successfully controlled by a modern medical treatment, and he was discharged from the hospital. After that episode, he continued to recover even in his terminal state of liver cirrhosis, and he survived until 2010. Of course, he never forgot to take 2000 mg of vitamin C at every meal (three times per day, for a total amount per day of 6000 mg) during his last 22 years. This episode motivated my study of the effects of vitamin C on liver damage. In fact, I have become familiar with an appropriate animal model for in vivo studies of vitamin C, which is the G*ulo*(−/−) mice model. These mice cannot synthesize vitamin C in their body because one oxidase-encoding gene involved in the terminal stage of vitamin C synthesis has been knocked out. Using this animal model, an experiment to confirm that vitamin C insufficiency could be a major risk factor for the development of liver disease was performed by my vitamin C research team. Furthermore, its specific effects and related mechanisms in vivo have been largely elucidated. When liver injury was induced in *Gulo*(−/−) mice by an injection of concanavalin A (Con A), greater extensive liver damage was observed, accompanied by an increased number of apoptotic hepatocytes in *Gulo*(−/−) mice with insufficient vitamin C. Additionally, the plasma and hepatic levels of the pro-inflammatory cytokines, in this case TNF-α and IFN-γ, were much higher in the *Gulo*(−/−) mice than in a control mouse group (Fig. 10.2).

Moreover, the increased numbers of liver-infiltrating T cells found in the *Gulo*(−/−) mice were related to the increased hepatic levels of IFN-inducible factor (IP-10) (Fig. 10.3). Although the *Gulo*(−/−) mice had higher amounts of interleukin-22 (IL-22), a hepatoprotective cytokine, defects in IL-22Ra expression and its downstream STAT3 activation in hepatocytes were also found (Fig. 10.4).

These results suggest that severe liver damage induced by inflammation could be prevented by sufficient supplementation with vitamin C. I think this report should be the first scientific document to demonstrate the novel in vivo action mechanisms of vitamin C with regard to preventing disease development in the liver through the regulation of excessive immune activation and the maintenance of the IL-22Ra signaling pathways (Bae et al. 2013).

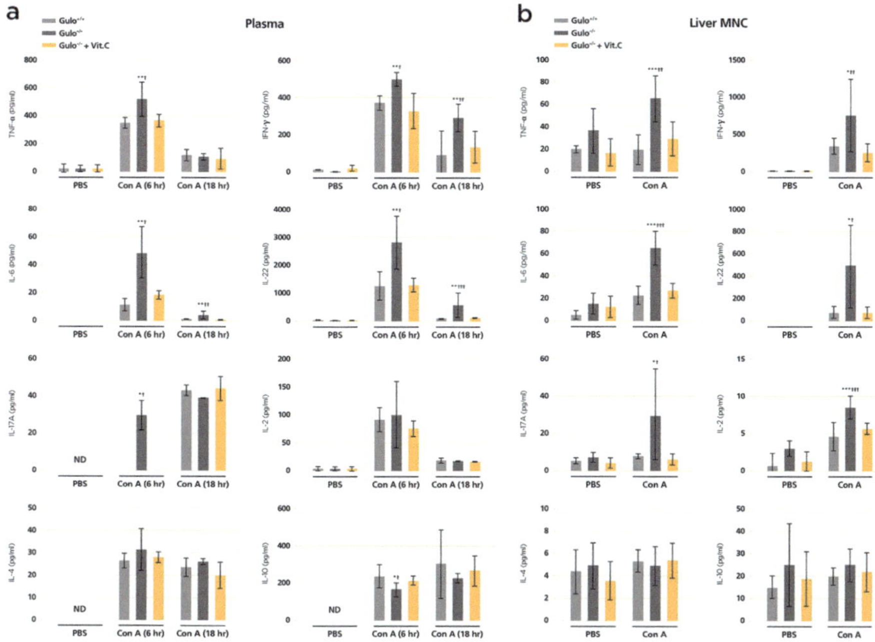

Fig. 10.2 The excessive production of pro-inflammatory cytokines in vitamin C-insufficient Gulo(−/−) mice in response to Con A. WT [Gulo(+/+)], vitamin C-insufficient [Gulo(−/−)], and vitamin C-sufficient [Gulo(−/−) + VC] mice were intravenously injected with 20 μg/g of Con A or PBS. (**a**) Increased cytokine levels in plasma. Vitamin C deficiency induced marked increase of pro-inflammatory cytokines in plasma by Con A injection, especially at the initial time (6 h after injection). (**b**) Increase of cytokine production from liver MNCs by Con A injection. Vitamin C deficiency also induced marked increase of pro-inflammatory cytokines in liver mononuclear cells by Con A injection, especially at the later time (18 h after injection) (Bae et al. 2013)

Our research team has also reported a very important clue related to the pathogenesis of liver cirrhosis using *Gulo*(−/−) mice, deficient in vitamin C, in which hepatotoxicity was induced by a thioacetamide (TAA) injection. They showed a decreased survival rate and increased serum markers for liver cell damage after the TAA injection (Fig. 10.5).

In addition, the levels of ROS and lipid peroxides in their liver were increased. Critically, the histological preparations clearly showed that collagen fiber depositions were remarkable (Fig. 10.6) and the number of the activated hepatic stellate cells increased. In an experiment using an immortalized human hepatic stellate cell line and primary rat stellate cells, vitamin C suppressed both their proliferation (Fig. 10.7) and hydrogen peroxide-induced collagen expression (Fig. 10.8), suggesting that vitamin C can prevent hepatic fibrosis by sufficiently scavenging ROS (Kim 2014).

Recently, an additional important report on the hepatoprotective effects of vitamin C was published after collaborative research between our team and the clinical hepatologists who had actually taken part in the medical care of my father-in-law,

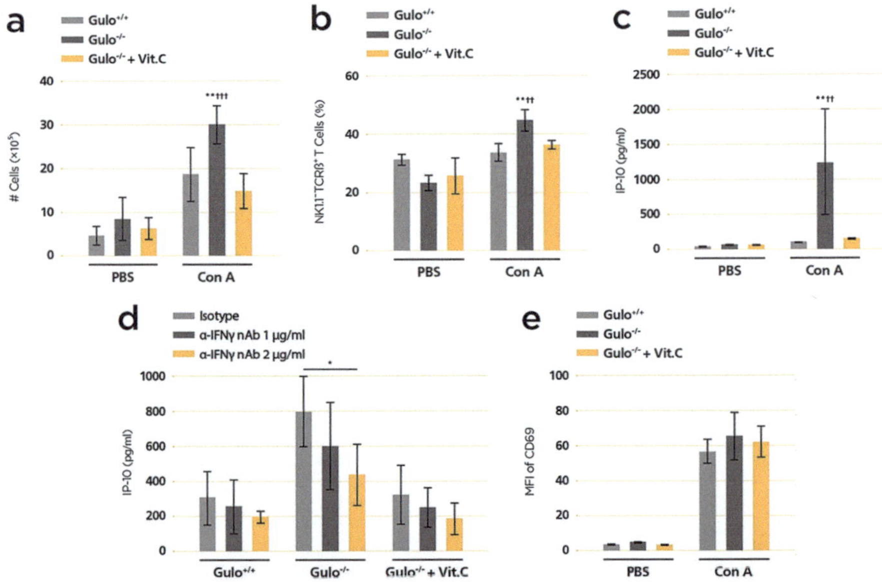

Fig. 10.3 Increasing numbers of liver-infiltrating T cells in vitamin C-insufficient Gulo(−/−) mice in response to Con A. After WT [Gulo(+/+)], vitamin C-insufficient [Gulo(−/−)], and vitamin C-sufficient [Gulo(−/−) +VC] mice were intravenously injected with 20 μg/g of Con A or phosphate-buffered saline (PBS), liver MNCs were isolated at different time points. (**a**) Increase of the total number of liver MNCs 6 h after Con A injection. Its increase was most prominent in vitamin C-insufficient Gulo(−/−) mice. (**b**) Increase of the T-cell frequency in liver MNCs by Con A injection. Flow cytometric analysis showed vitamin C deficiency induced an increase of T cell. (**c**) Increased IP-10 production from liver MNCs by Con A injection was most conspicuous in vitamin C-insufficient Gulo(−/−) mice. (**d**) Suppression of IP-10 production by the neutralization of IFN-γ bioactivity. (**e**) The equal activation status of liver MNCs was identified by surface expression of CD69 (Bae et al. 2013)

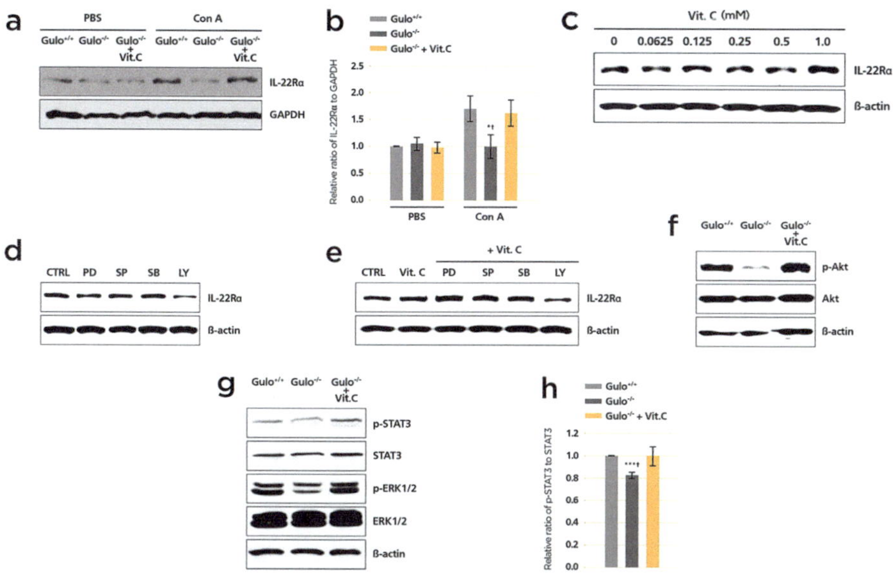

Fig. 10.4 Defects of PI3K/Akt-dependent IL-22Ra expression and its downstream STAT3 activation in the hepatocyte of vitamin C-insufficient Gulo(−/−) mice. (**a**) The Western blot analysis showed defect on IL-22Ra expression in the liver of vitamin C-insufficient Gulo(−/−) mice. (**b**) The ratio of IL-22Ra/GAPDH was lowest in vitamin C-insufficient Gulo(−/−) mice. (**c**) IL-22Ra expression by the Western blot analysis on the murine hepatocyte cell line, FL83B, was increased in a dose-dependent manner of vitamin C. (**d**, **e**) Only PI3K/Akt-dependent IL-22Ra expression (specific inhibitor; LY294002) was inhibited on the murine hepatocyte cell line, when FL83B was pretreated with PD98059 (20 μM), SP600125 (10 μM), SB203580 (20 μM), and LY294002 (5 μM) for 1 h and then incubated for another 24 h in the absence (**d**) or presence (**e**) of 1 mM of vitamin C, and then IL-22Rα and β-actin expression was determined by the Western blot analysis. (**f**) Defect on the phosphorylation of Akt in hepatocytes upon Con A stimulation was observed in vitamin C-insufficient Gulo(−/−) mice. (**g**) Defect on the activation of STAT3 and ERK1/2 in hepatocytes of vitamin C-insufficient Gulo(−/−) mice upon Con A stimulation was observed in vitamin C-insufficient Gulo(−/−) mice. (**h**) The ratio phosphorylated STAT3/total STAT3 was lowest in vitamin C-insufficient Gulo(−/−) mice. (**i**) Defect on STAT3 activation in hepatocytes from vitamin C-insufficient Gulo(−/−) mice by recombinant IL-22 stimulation was observed. *ERK* extracellular signal-regulated kinases, *IL-22* interleukin-22, *STAT* signal transducers and activators of transcription (Bae et al. 2013)

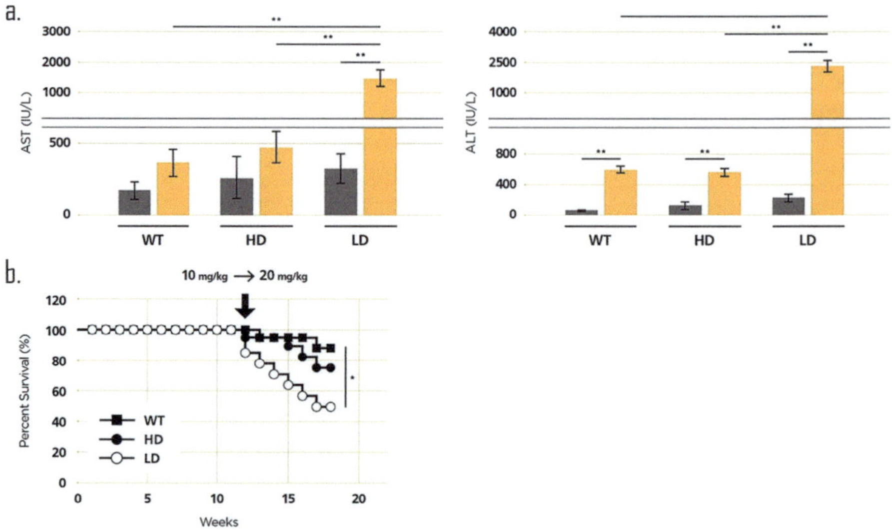

Fig. 10.5 Vitamin C insufficiency exacerbated TAA-induced hepatotoxicity. (**a**) Hepatotoxicity, as indicated by the serum levels of AST (left) and ALT (right) activities before (black bars) and after (yellow bars) TAA treatments. Serum levels of AST and ALT were markedly increased in vitamin C-deficient mice by TAA treatment. (**b**) Survival rates were exacerbated in the LD group (vitamin C-deficient mice). *AST* aspartate aminotransferase, *ALT* alanine aminotransferase, *HD* high-dose group, *LD* low-dose group, *WT* wild-type group (Kim J et al. 2014)

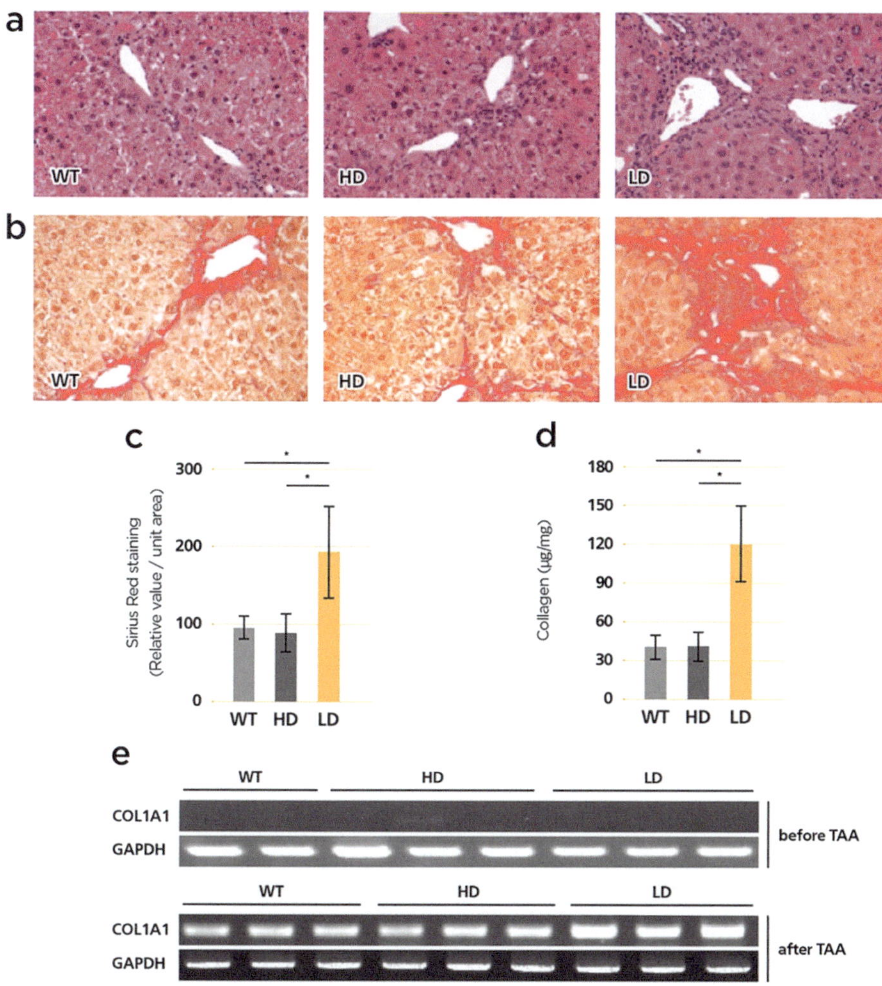

Fig. 10.6 Vitamin C deficiency aggravates liver fibrosis of mice. (**a**) In hematoxylin and eosin staining of the TAA-treated liver tissue, fibrotic tissue change around the periportal regions was conspicuous in the vitamin C-deficient mice liver. (**b**) Under Sirius red staining, collagen fibers were prominent in the LD group (vitamin C-deficient mice) (**c**) and (**d**). The biochemically measured amount of collagen was drastically increased only in the LD group by TAA treatment (**e**). RT-PCR showed an increase of expression collagen encoding gene in the LD group by TAA treatment; *HD* high-dose group, *LD* low-dose group, *WT* wild-type group, *TAA* thioacetamide

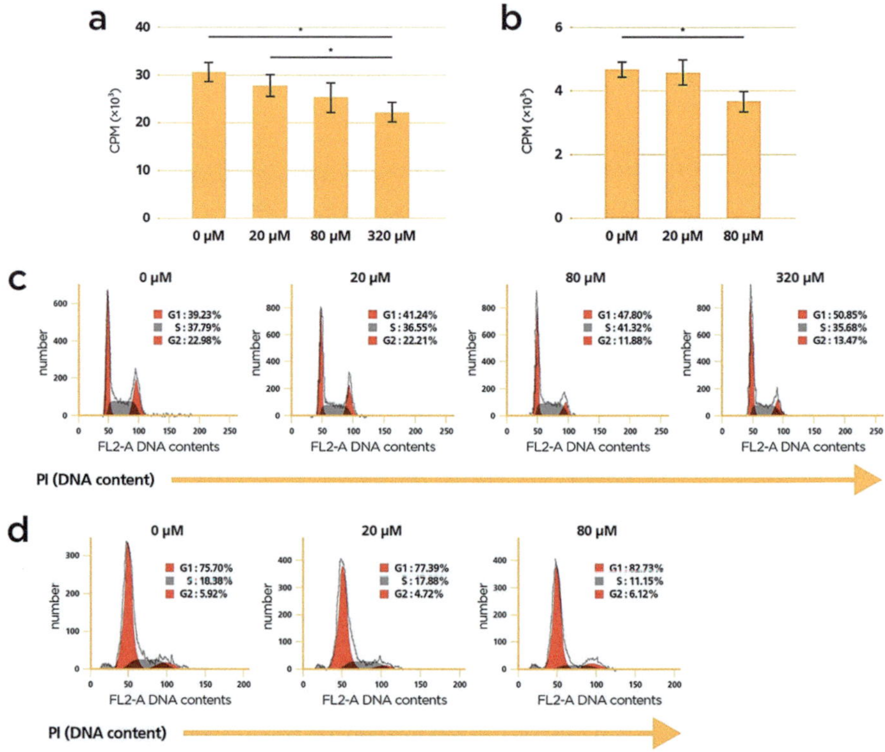

Fig. 10.7 Vitamin C suppressed HSC proliferation by cell cycle arrest. LX-2 cells and the rat primary HSCs were cultured in the presence of vitamin C at concentrations indicated in the figures. (**a**) Thymidine uptake of LX-2 cells and (**b**) the rat primary HSCs were decreased by in vitro application of vitamin C dose-dependently. The frequency of LX-2 cells (**c**) and the rat primary HSCs (**d**) were increased in the G1phase by vitamin C treatment, indicating cell cycle arrest. *HSCs* rat primary hepatic stellate cells, *LX-2 cells* immortalized human hepatic stellate cells, *PI* propidium iodide (Kim et al. 2014)

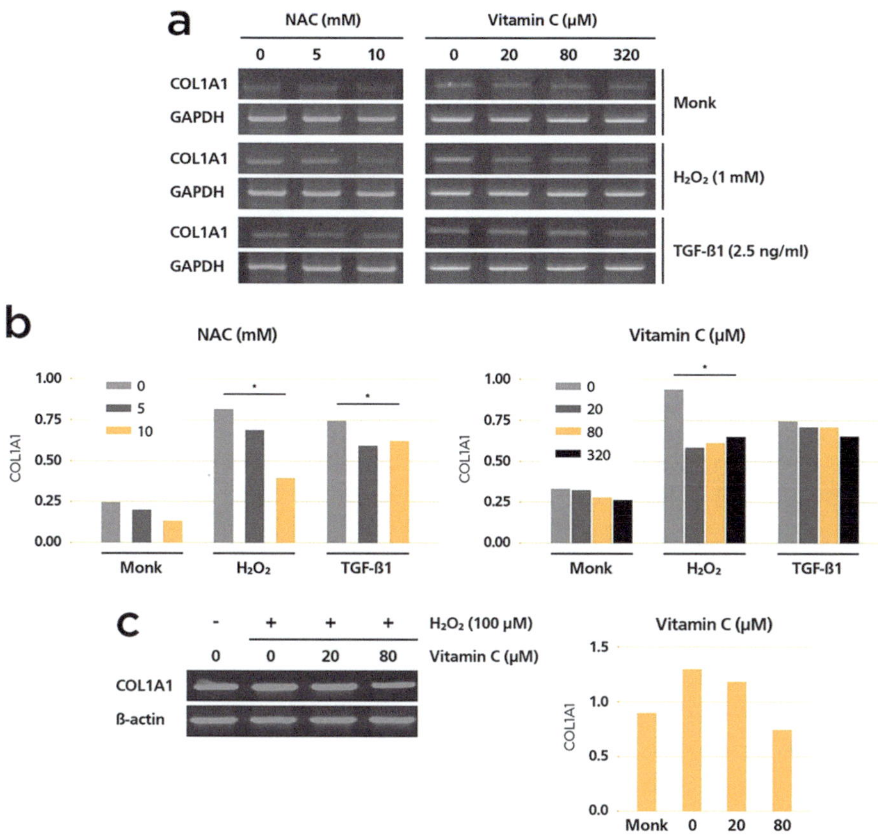

Fig. 10.8 Vitamin C suppressed H2O2-induced COL1A1 expression. LX-2 cells were serum-starved for 48 h and then treated with H2O2 or TGF-β1 to induce collagen expression, in the presence of NAC or vitamin C at various concentrations. (**a**) RT-PCR analysis for the cells at 24 h after treatment and (**b**) the ratio of COL1A1/GAPDH showed that with H2O2 or TGF-β1 treatment, LX-2 cells increased the transcriptional level of col1a1 and that whereas NAC suppressed both H2O2- and TGF-β1-induced COL1A1 expression, vitamin C suppressed only the expression induced by H2O2, not that by TGF-β1. (**c**) When the rat primary HSCs were treated with H2O2, collagen expression was induced, which was decreased in the presence of vitamin C dose-dependently. *NAC* N-acetylcysteine (Kim et al. 2014)

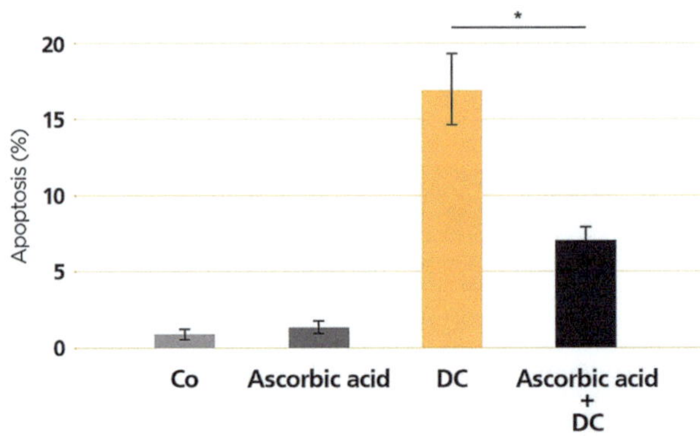

Fig. 10.9 Bile acid-induced hepatocyte apoptosis and its attenuation by vitamin C. Apoptotic Huh-BAT cells were increased by treatment of DC (200 μM) for 4 h in the absence of vitamin C (1 mM), but their apoptosis was greatly inhibited by treatment of vitamin C (1 mM). Apoptotic cells were counted by DAPI staining and fluorescent microscopy. *DC* deoxycholate (Yu et al. 2015)

who long suffered from liver cirrhosis, as mentioned above. They observed the effects of vitamin C on lithocholic acid (LCA)-induced liver injury in *Gulo(−/−)* mice. One of their important findings was that vitamin C attenuated bile acid-induced hepatocyte apoptosis in vitro (Fig. 10.9).

Furthermore, oral intake of LCA increased plasma cholestasis markers and resulted in more extensive liver injury and hepatic fibrosis caused by more prominent apoptotic cell death and the recruitment of more intrahepatic inflammatory CD11b+ cells into the liver of the *Gulo*(−/−) mice (Fig. 10.10). They suggested from these results that vitamin C could have a hepatoprotective effect against cholestatic liver injury (Yu et al. 2015).

In addition to these reports, several papers suggesting hepatoprotective effects of vitamin C on liver tissue damaged by chemicals such as CCl4 (Su et al. 2014) and monosodium glutamate (MSG) (El-Meghawry El-Kenawy et al. 2013) have been published.

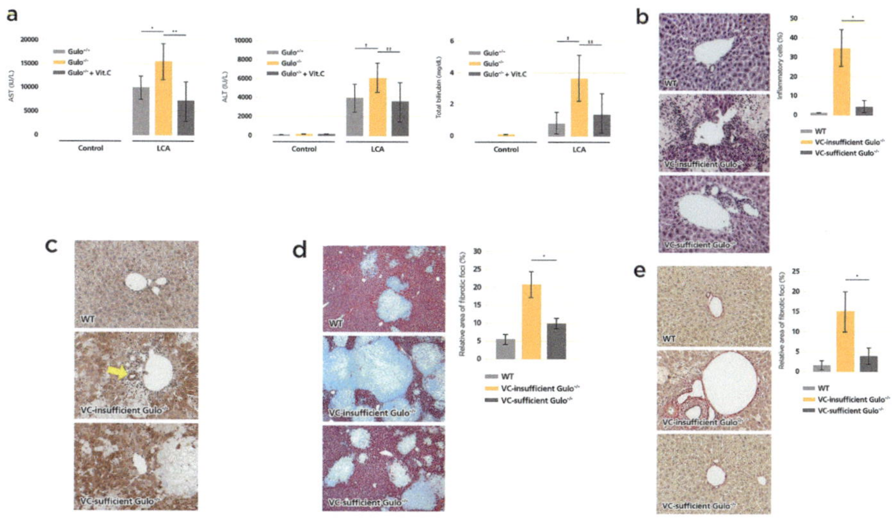

Fig. 10.10 The extensive liver damage and hepatic fibrosis in vitamin C-insufficient Gulo(−/−) mice in response to LCA. (**a**) The plasma levels of AST, ALT, and total bilirubin were increased by LCA feeding (1%, 4 days). (**b**) H&E staining images showed that the extensive liver injury by LCA feeding (0.5%, 24 h) was observed only in vitamin C-deficient mice. (**c**) Cytokeratin 19 staining showed that there was a dense staining of bile duct only in vitamin C-deficient mice, indicating bile duct injury by LCA feeding (1%, 4 days). Yellow arrow indicated bile duct. (**d**) Masson's trichrome staining showed that extensive hepatic fibrosis (green color) by LCA feeding (1%, 4 days) was observed only in vitamin C-deficient mice. (**e**) The Pico-sirius red staining revealed more prominent peri-ductal fibrosis by LCA feeding (1%, 4 days) in vitamin C-insufficient Gulo(−/−) mice than in vitamin C-sufficient Gulo (**c**) mice. *LCA* lithocholic acid, *WT* wild type, *VC* vitamin C Yu et al. 2015)

References

Bae S, Cho CH, Kim H, Kim Y, Kim HR, Hwang YI, Yoon JH, Kang JS, Lee WJ. In vivo consequence of vitamin C insufficiency in liver injury vitamin C ameliorates T-cell-mediated acute liver injury in Gulo(−/−) mice. Antioxid Redox Signal. 2013;19(17):2040–53.

De Brito T, Shimizu SH, Yamashiro E, Da Silva LC. Chronic human schistosomiasis mansoni-schistosomal antigen, immunoglobulins and complement C3 detection in the liver. Liver. 1985;5(2):64–70.

El-Meghawry El-Kenawy A, Osman HE, Daghestani MH. The effect of vitamin C administration on monosodium glutamate induced liver injury; an experimental study. Exp Toxicol Pathol. 2013;65(5):513–21.

Kim JH, Jeong YJ, Hong JM, Kim HR, Kang JS, Lee WJ, Hwang YI. Chronic vitamin C insufficiency aggravated thioacetamide-induced liver fibrosis in gulo-knockout mice. Free Radic Biol Med. 2014;67:81–90.

Su M, Chen H, Wei C, Chen N, Wu W. Potential protection of vitamin C against liver-lesioned mice. Int Immunopharmacol. 2014;22(2):492–7.

Trevisan A, Cavigli R, Meliconi R, Stefanini GF, Zotti S, Rugge M, Noventa F, Betterie C, Realdi G. Detection of immunoglobulins G and a on the cell membrane of hepatocytes from patients with alcoholic liver disease. J Clin Pathol. 1983;36(5):530–4.

Yu SJ, Bae S, Kang JS, Yoon JH, Cho EJ, Lee JH, Kim YJ, Lee WJ, Kim CY, Lee HS. Hepatoprotective effect of vitamin C on lithocholic acid-induced cholestatic liver injury in Gulo(−/−) mice. Eur J Pharmacol. 2015;5(762):247–55.

Chapter 11
Skin

Structures of the Skin

In adults, the skin is the heaviest and largest as a single organ in the body. Actually, its proportion to the total body weight reaches 15–17%. In reality, it is a symbol of beauty, especially for women. Therefore, most people, especially all women, are usually doing their best to keep their skin healthy and clear. Several conditions of the skin, such as the wrinkles, tightness, the extent of pigmentation, and so on, are seen as landmarks of aging.

The skin consists of three layers: epidermis, dermis, and hypodermis. The epidermis is composed mainly of typical skin cells, keratinocytes. However, it also contains some other cells although found in small numbers including the following: (1) melanocytes, which produce melanin pigments, playing a key role in determining skin color; (2) Langerhans cells, which are bone marrow-derived macrophages functioning as immune cells of the skin, specifically binding to antigens exposed to the skin and processing and presenting them to T lymphocytes; and (3) Merkel's cells, which have been thought to be a sensory mechanoreceptor because of the free nerve endings, frequently found at the base of these cells.

The keratinocytes, which mostly make up the epidermis and produce keratin, are the representative cells of several adult stem cells because they keep proliferating without stopping during people's lifetimes, forming five layers of the epidermis (from bottom to top: the basal layer, spinous layer, granular layer, clear layer, and corneal layer) according to the characteristics of the keratinocytes comprising each layer.

There are two kinds of skin, thick and thin ones, according to thickness of the epidermal layer. The thick skin, which is usually non-hairy, is found on the palms and soles, and all other areas of the human body have a thin skin, which is usually hairy.

W. J. Lee, *Vitamin C in Human Health and Disease*,
https://doi.org/10.1007/978-94-024-1713-5_11

Characteristically, the epidermis has no blood vessels so nutrition is supplied by simple diffusion from abundant vessels in the dermis located just beneath the epidermis.

The dermis, which is located beneath the epidermis, is rich in blood vessels and has abundant fibers, which consist of collagenous fiber, elastic fiber, and reticular fiber, giving the skin elasticity and tightness. It is known that a deficiency in collagen fibers due to aging causes wrinkles. The dermis contains extensive microvascular networks as follows: (1) a rich capillary network is located over the superficial layer of the dermis, especially around the epidermal ridges, playing a key role in the regulation of the core body temperature and nourishment of the overlying epidermis due to a lack of blood vessels, and (2) arteriovenous shunts are found in some skin regions, through which blood can pass directly from the arteries to the veins functioning to regulate the blood pressure and body temperature. The proportion of blood contained in the dermal blood vessels is approximately 4–5% of the total blood in the body.

In the dermis, most of all sensory nerve endings are present so that almost all sensations experienced in our lives, such as pain, touch, pressure, and temperature, can be felt through these nerve endings in the dermis. Even the postganglionic fibers of the sympathetic nerve are distributed around the sweat glands and the hair follicles located in the dermis.

As mentioned above, the dermis contains epidermal derivatives, such as the hair follicles and sweat and sebaceous glands.

The hypodermis is also called subcutaneous tissue, located just beneath the dermis, connecting the skin to the underlying organs, which are usually the skeletal muscles. It consists of loose connective tissues, especially fat tissue, which is comprised of many fat cells. The thickness is determined by the number of fat cells, which is different, according to the areas of the body. Actually, it covers the underlying skeletal muscles, which is a reason why it is also called superficial fascia (Fig. 11.1).

Functions of the Skin

The skin has many functions. First, it functions as a very important primary defense system against all kinds of foreign materials, including any microorganisms, toxic substances, and even simple water, for which the junctional complexes consisting of adhering junction, tight junction, and gap junctions are installed between the keratinocytes. A special protection by the skin from ultraviolet (UV, wavelength 290–320 nm) rays is carried out by melanocytes, which are located in the bottom of the epidermis, the basal layer, and produce melanin pigments. Actually, the melanin pigments synthesized in the melanocytes are transferred to the keratinocytes consisting of the basal and spinous layers and eventually located at the supranuclear region of the cytoplasm, thus preventing the nuclei from receiving the harmful effects of sunlight. In humans, the ratio of melanocytes to keratinocytes in the basal

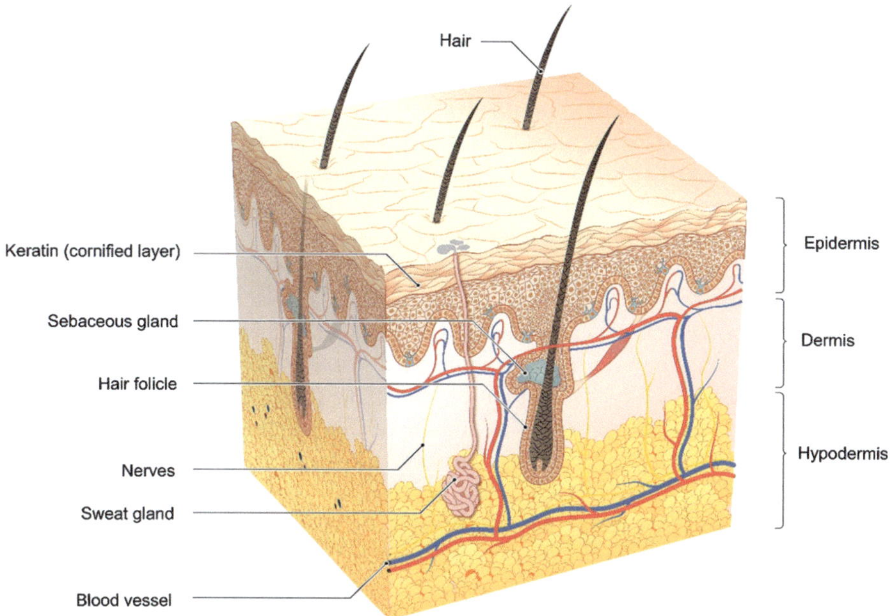

Fig. 11.1 Structures of the skin. The skin consists of three layers, epidermis, dermis, and hypodermis, in the order of outside to inside

layer is constant within each area of the body, but there are some differences from one region to another. Sex or race looks like to have no influence in determining the number of melanocytes. Differences in skin color among the races are determined by the difference in the number of melanin granules in the keratinocytes. However, it seems that the regional environment may have some roles in determining the skin color. One good example is that African people who live nearer to the equatorial line have darker skin than other people who do live far from the equatorial line.

The skin also has a function regulating the body temperature. There is a thermoregulation center in the hypothalamus, by which the body temperature of humans is controlled automatically. When the body temperature is above the core body temperature (37.5 °C) set in the hypothalamus, the cooling system in the body is working, while when it is below 37.5 °C, the heating system is working. The heating system is comprised of (1) muscle contractions, (2) erection of skin hair, and (3) vasoconstriction. In reality, when people feel cold, first, all the hairs on the skin stand up, which is the so-called goose bumps, for which all arrector pili muscles attached to the shaft of the hair and located in the upper dermis of the skin distributed over all areas of the body except for the soles and palms contract, thus generating heat. Furthermore, because the standing hairs can contain much more air than bending hairs, the loss of heat through the skin from the body can be effectively reduced. In addition, the cold induces vasoconstriction of all small blood vessels abundantly present in the dermis to reduce the heat loss through the skin. In contrast to this, the cooling system of the body is working. When people feel hot, first, all small blood vessels in the dermis are vasodilated, thus facilitating release of heat from the body to cool down the elevated body temperature. Next, the sweat glands located in the deep dermis of the skin over the body produce the appropriate amount of sweat, according to the elevated body temperature. It is well known that 539 kcal of energy are needed to evaporate 1 ml of pure water. Eventually, the core body temperature can be kept constant by the united efforts of these heat controlling apparatuses, most of which are located in the skin.

Another function of the skin is to mediate the synthesis of the active form of vitamin D. In keratinocytes, 7-dehydrocholesterol is converted to precholecalciferol by UV radiation, which is isomerized into cholecalciferol. In the liver, cholecalciferol is hydroxylated into 25-hydroxycholecalciferol, which is finally hydroxylated to become 1,25-dihydroxycholecalciferol, the active form of vitamin D, in the kidney, which has a crucial role in increasing the calcium uptake in the intestine. This conversion of cholecalciferol into 25-hydroxycholecalciferol and 1,25-dihydroxycholecalciferol also takes place in the skin. It is very important to know that vitamin D synthesized in the skin can be used more effectively than vitamin D absorbed by the diet because the former can rarely be inactivated in the liver, while the latter can easily be inactivated in the liver. Therefore, the importance of sunlight exposure to the appropriate extent during daily life for your health cannot be overemphasized, although every dermatologist used to say that the greatest enemy to skin health must be sunlight.

Role of Vitamin C in Skin Health

So many good effects of vitamin C on skin health have been reported such that most cosmetics commercialized for skin health and beauty are advertised to contain lots of vitamin C. However, considering the biochemical properties of vitamin C, one should be very skeptical about the contents of these advertisements which most cosmetic companies use. First, vitamin C is an acid, ascorbic acid, making it hydrophilic. However, the epidermis is usually not only covered by some lipids, sebum, secreted by the sebaceous glands located at the upper dermis, around the shaft of the hair follicles. The cell membranes of the keratinocytes consisting of the epidermis are also hydrophobic. Therefore, even though much vitamin C is contained in several commercial products, actually that vitamin C in the form of ointments or creams put on the skin is unable to penetrate and invade into the skin at all. Even worse, the vitamin C is easily oxidized in the air and has no effects during its stay on the skin for penetration and invasion. That is the reason why many commercial companies are making every effort to develop methods on how to penetrate vitamin C into the skin successfully. Some companies are trying to make ultra-micropores on the skin without apparent skin damages, and some companies are utilizing nano-vehicles containing vitamin C, which can transport vitamin C into the epidermis without its oxidation. Now, I do not think there are successful commercial products using vitamin C for skin care. Anyhow, despite that, vitamin C has definite beneficial effects on the skin when it can be applied to the skin cells successfully via any route. The summarized effects of vitamin C on the skin are as follows: (1) whitening effect, (2) delay of wrinkle formation, and (3) protection from UV radiation.

Whitening Effects

Most women have endless envy for white and clean skin. There are several factors in determining skin color, such as skin pigments (melanin and carotene), the number of blood vessels in the dermis, and the color of the blood flowing in the vessels of the dermis, and the most important one is the contents of the skin pigments. Actually, skin color is decided by the content of the melanin pigments. Synthesis of melanin is initiated by tyrosine, which is transformed into 3,4-dihydroxyphenylalanine (DOPA) and then into dopaquinone, which is finally converted into melanin. In this process, an enzyme, tyrosinase, has a key role. Some Japanese dental scientists have reported that vitamin C significantly inhibited tyrosinase activity and melanin formation in B16 mouse melanoma cells. Even they showed that vitamin C could have inhibitory effects on melanin formation using a three-dimensional human skin model (Shimada et al. 2009). However, it does not seem that the level of vitamin C in the plasma, skin, and urine has a correlation with the pathogenesis of depigmentation skin diseases such as albinism and vitiligo (Ratnam et al. 1977).

Delay of Wrinkle Formation

One of the landmarks for aging is wrinkle formation in the skin. Generally, wrinkles are formed in the area where habitual folding takes place. For example, wrinkles are rarely found on the cheek, where no habitual folding takes place, while they are frequently found on the forehead of so many people, where physical folding frequently takes place whenever people raise their eyes. However, with age, they appear even in rare folding areas such as the cheek. It has been known that a lack of collagen synthesis in the dermis with aging makes a shallow groove on the skin, the so-called wrinkle. Historically the best well-known scientific fact of vitamin C is that it is a very important cofactor involved in the process of collagen biosynthesis, which has been written down in a textbook about medicine; thus, there is even no need to refer to scientific papers for evidence on the role of vitamin C in the collagen biosynthesis pathway. Vitamin C always exerts its effect chronically. A persistent supply of a sufficient amount of vitamin C can definitely delay wrinkle formation by inhibiting the decreases in collagen biosynthesis, which is an unavoidable phenomenon in old people.

Protection from UV Radiation

Although UV radiation has several beneficial biological functions such as vitamin D biosynthesis, germicidal activity, and so on, cosmetically, it is considered as a public enemy for everyone, especially for all women, regardless of their age, because it is well known that it could cause chronic skin damage and induce skin aging eventually (Ichihashi et al. 2003). It is also known that there are two ways in induction of skin damage by UV radiation: one is UV-induced direct damage and the other is UV-induced generation of reactive oxygen species (ROS) in cells, which increases the pro-inflammatory cytokines (IL-1, IL-6, and IL-18) and induces skin inflammation, which eventually leads to skin aging (Scharffetter-Kochanek et al. 1997; de Gruijl et al. 2001).

There are two ways to protect the skin from UV-induced damage. One is chemical protection, and the other is physical protection, which is very simple and can be done simply by blocking UV ray penetration or preventing UV exposure physically such as wearing a mask, shielding sunlight by an umbrella, wearing a cap, etc. Chemical protection works by removing the ROS generated by UV radiation in the epidermis. Vitamin C is a representative antioxidant to protect the skin from the inflammatory damages induced by UV radiation. Our research team has revealed a specific mechanism by which vitamin C has a protective effect on the skin against injuries from UV radiation. According to their report, UV radiation induces inflammation in the skin by releasing IL-18 and monocyte chemoattractant protein (MCP)-1 (Fig. 11.2), which are suppressed by vitamin C treatment (Fig. 11.3). Very

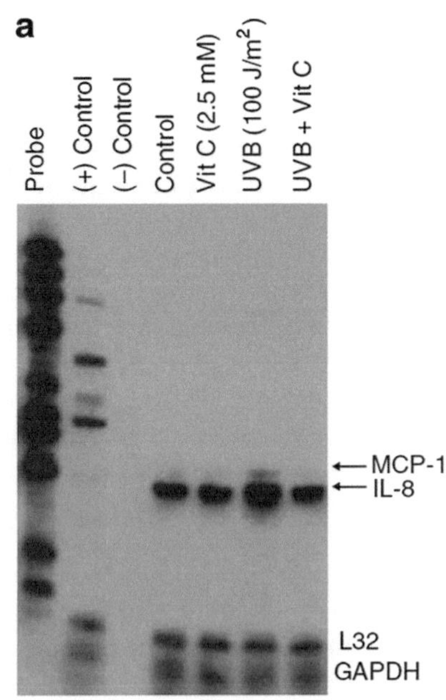

Fig. 11.2 RNAse protection assay for MCP-1 and IL-18 in the human keratinocytes after UV irradiation. This figure shows that UVB irradiation to the human keratinocytes induces release of IL-18 and MCP-1 and their production is decreased by vitamin C treatment. (Kang et al. 2007)

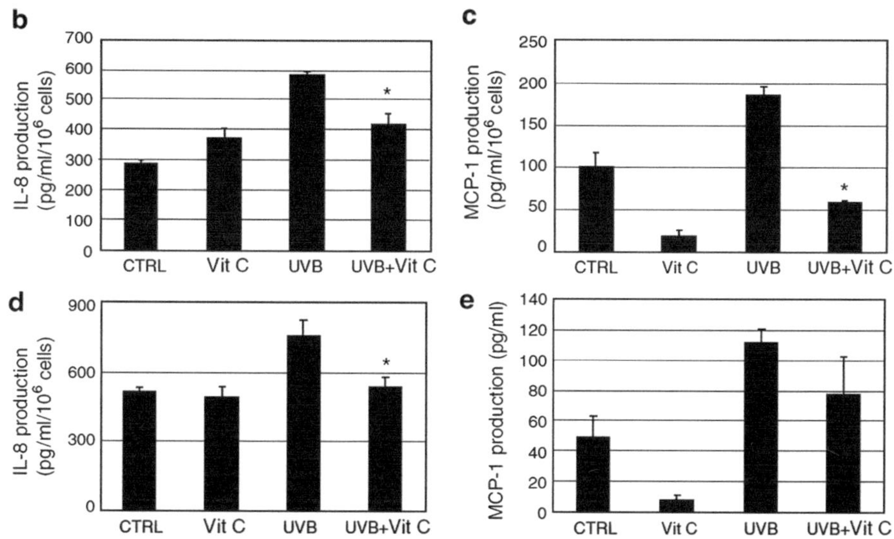

Fig. 11.3 ELISA assay for MCP-1 and IL-18 in the human keratinocytes after UVB irradiation. Increased productions of IL-18 and MCP-1 by UVB irradiation are suppressed by vitamin C treatment. (Kang et al. 2007)

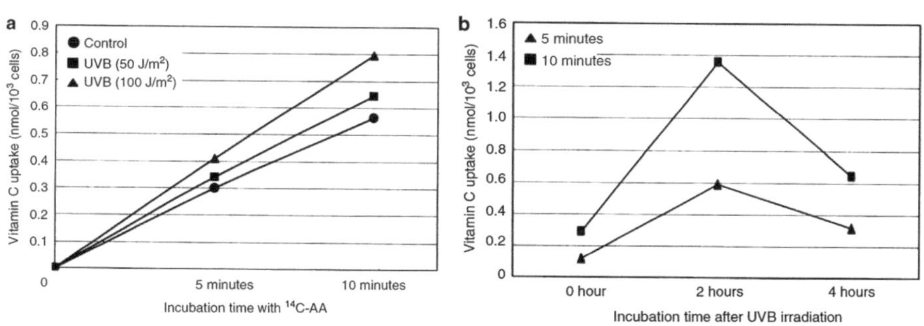

Fig. 11.4 Measurement of vitamin C uptake in the keratinocytes after UVB irradiation. Vitamin C uptake increases in a time-dependent and a dose-dependent manner by UVB irradiation. (Kang et al. 2007)

interestingly, vitamin C uptake into keratinocytes is increased by UVB irradiation in a time- and dose-dependent manner (Fig. 11.4) through translocation of sodium-dependent vitamin C transporter-1 (SVCT1), a vitamin C-specific transporter, from the cytosol to the membrane (Fig. 11.5) (Kang et al. 2007).

However, there are practical problems in the preparation of vitamin C products in the form of ointments for skin protection from UV irradiation. Consider the fact that vitamin C in the body can only be supplied to the dermis and not the epidermis because there are no blood vessels in the epidermis including capillaries. Therefore, vitamin C products for skin care should be spread onto the skin surface and should

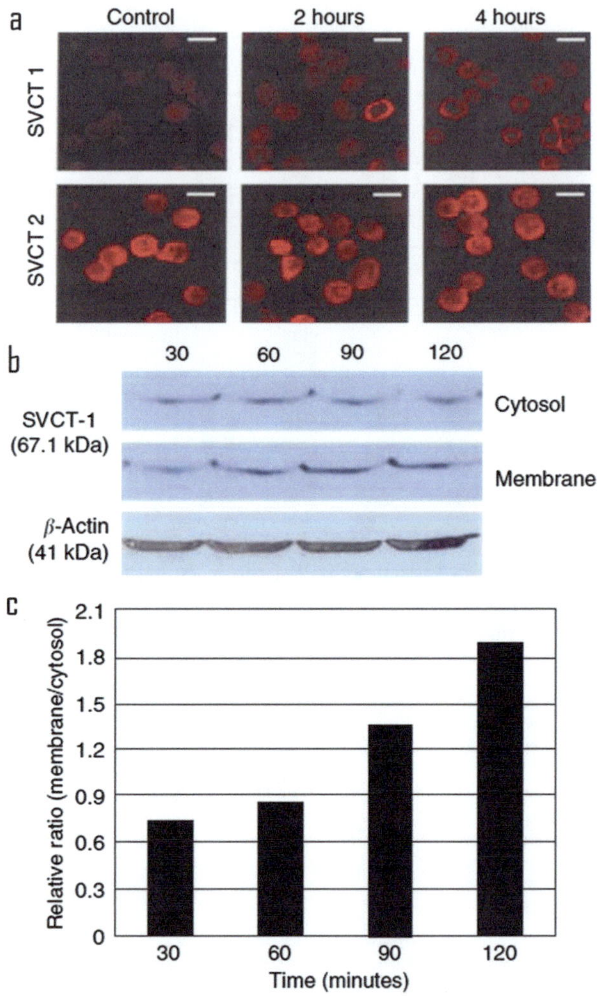

Fig. 11.5 Translocation of SVCT1 from the cytosol to the membrane. (**a**) Fluorescent staining of keratinocytes for SVCT1 and 2. Vitamin C treatment for keratinocytes after UV irradiation induces change in distribution pattern of SVCT1, but not of SVCT2. (**b**) Western blotting of subfractions of cells for SVCT1 after treatment of vitamin C for UV-irradiated keratinocytes. (**c**) Increase of the ratio (membrane/cytosol) of SVCT1 after vitamin C treatment for UV-irradiated keratinocytes indicates that SVCT1 is translocated from the cytosol to the membrane. (Kang et al. 2007)

penetrate to the base of the epidermis without any chemical decomposition. In reality, vitamin C is ascorbic acid, which has a negative charge and is usually hydrophilic. As such, it is very hard for vitamin C to penetrate through several layers of the epidermis, of which the surface is usually hydrophobic. In addition, vitamin C is easily oxidized in ambient condition and decomposed while on the surface of the skin before penetration. Therefore, many scientists trying to commercialize vitamin C products for skin care have done their best to develop stable products without chemical change, such as ascorbic monophosphate, ascorbic diphosphate, or ascorbic triphosphate, which do not exert the original functions of vitamin C. Nowadays, a stable form of vitamin C has been developed and used for commercial products while keeping the original functions of the native form of vitamin C without any chemical deposition on the skin surface.

References

De Gruijl FR, van Kranen HJ, Mullenders LH. UV-induced DNA damage, repair, mutations and oncogenic pathways in skin cancer. J Photochem Photobiol B. 2001;63:19–27.

Ichihashi M, Ueda M, Budiyanto A, et al. UV-induced skin damage. Toxicology. 2003;189:21–39.

Kang JS, Kim HN, Jung DJ, Kim JE, Mun GH, Kim YS, Chung JH, Park KC, Cho D, Shin DH, Hwang YI, Lee WJ. Regulation of UVB-induced IL-8 and MCP-1 production in keratinocytes by increasing vitamin C uptake via the redistribution of SVCT-1 from the cytosol to the membrane. J Invest Dermatol. 2007;127:698–706.

Ratnam AV, Sastry PB, Satyanarayana BV. Ascorbic acid and melanogenesis. Br J Dermatol. 1977;97(2):201–4.

Scharffetter-Kochanek K, Wlaschek M, Brenneisen P, et al. UV-induced reactive oxygen species in photocarcinogenesis and photoaging. Biol Chem. 1997;378:1247–57.

Shimada Y, Tai H, Tanaka A, Ikezawa-Suzuki I, Takagi K, Yoshida Y, Yoshie H. Effects of ascorbic acid on gingival melanin pigmentation in vitro and in vivo. J Periodontol. 2009;80(2):317–23.

Chapter 12
Eye

The eye is a ball-like structure for vision which is a complicated and well-developed photosensitive organ. The eyes are located in the orbits, which are parts of the skull composing the face and are global bony spaces. Their functions include all activities related to vision, which is to discriminate the form of each object and to sense the color and light intensity reflected from some objects. They are constructed for their functions and are therefore composed of three concentric layers: (1) an outer layer, of which the main function is to support the eye structure, and which consists of the sclera and the cornea; (2) a middle layer, of which the function is to nourishing the structure, also known as the vascular layer and consisting of the choroid, ciliary body, and the iris; and (3) an inner layer, a type of nervous tissue which performs the proper function of the eyes and which is referred to as the retina, consisting of the inner retina proper and the outer pigment epithelium. The eyeballs are filled with a transparent gel, also called the vitreous body, consisting of 99% of water, a very small proportion of collagen, and heavily hydrated glycosaminoglycan containing hyaluronic acid as the main component (Fig. 12.1).

Function of Vision

For vision, light should be transmitted into the eye and finally reach photoreceptor cells, where the sense of vision is accomplished. The resulting signal is sent to the visual cortex through the optic nerve connected to the posterior portion of the eyeballs. For visual function, the light must pass through several refractive media. First, the light goes through the cornea, which is colorless and transparent but consists of five layers. The cornea has no blood vessels but is supplied with the richest network of sensory nerves among other eye tissues. Next, the light passes through the aqueous humor, contained in the anterior and posterior chambers, between which the iris is located. The iris is an apparatus which controls the amount of transmitted light. The aqueous humor has a composition similar to that of plasma but is a protein-poor

W. J. Lee, *Vitamin C in Human Health and Disease*,
https://doi.org/10.1007/978-94-024-1713-5_12

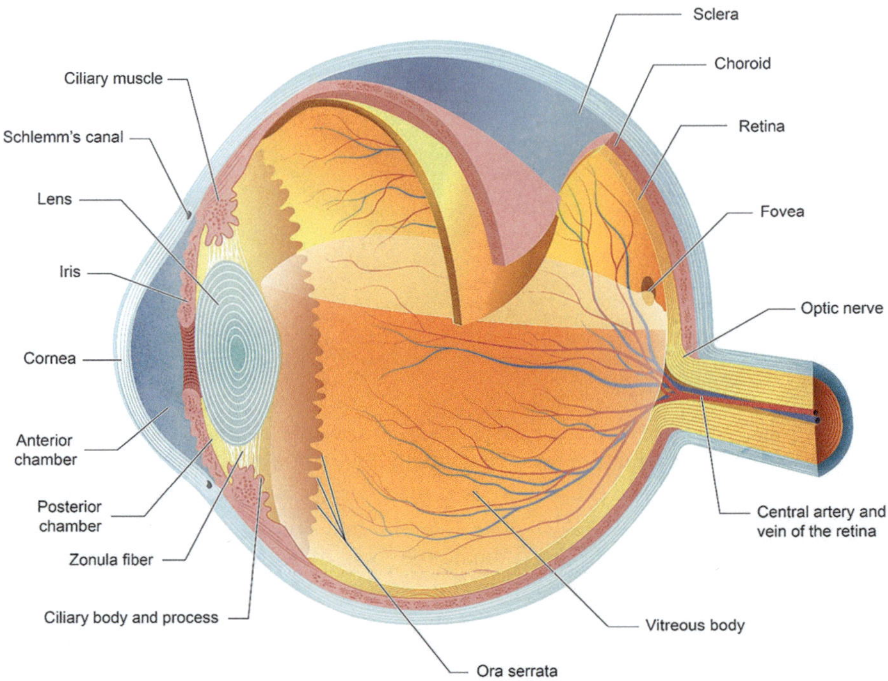

Fig. 12.1 Schematic drawing of eyeball. The wall of eyeball consists of three layers: (1) outer one, cornea and sclera; (2) middle one, iris, ciliary body in the front part, and choroid in the posterior part of the eyeball; (3) inner one, retina

clear fluid (0.1% protein vs plasma has 7% of protein). Moreover, it is produced in the epithelium lining the ciliary body and the processes of the posterior chamber. It passes through the cleft between the lens behind and the iris in front, entering the anterior chamber and finally draining into Schlemm's canal located in the corner between the cornea and the iris. The light passes through the lens, which is located between the chambers and the vitreous space filled with the transparent gel and connected to the ciliary process by the zonule on both sides, which enables the lens to fulfill the accommodation function. The accommodation function refers to control over near and distant vision. The lens belongs to one of the middle layer structures, in this case the choroid, ciliary body, and iris, considering its location. This biconvex structure has great elasticity and consists of three main components. One is the lens capsule, which covers the lens and essentially coats the outer surface of the underlying epithelium. It is homogenous, refractile, and rich in carbohydrates. The second component is the subcapsular epithelium, which is present only on the anterior surface of the lens. The final component is the lens fiber cells, highly differentiated cells deriving from the subcapsular epithelium which eventually lose their nuclei and other organelles. Crystalline proteins fill the spaces among these cells. The production of lens fiber cells gradually decreases with age.

The light transmitted through the lens continues through to the vitreous body filling the space in the eyeball, finally reaching the posterior pole of the eye ball, the retina, consisting of an outer pigment epithelium and the inner retina proper. The outer pigment epithelium consists of columnar cells, firmly attached to an underlying membrane. Junctional complexes are well developed between the columnar epithelial cells such that leakages of the light can be completely blocked for clear vision. In addition, the associated cytoplasm has a rich smooth endoplasmic reticulum, where vitamin A is esterified and transported to the photoreceptors. Apical cytoplasm and related microvilli also have numerous melanin granules, which are synthesized by a mechanism similar to that involved in melanin synthesis by melanocytes in the skin. On the inner side of the outer pigment epithelium layer, there are three inner layers of photosensitive cells. The first of these is an outer layer consisting of photosensitive cells, rods, and cones, which are the first photoreceptor cells to receive the light. The rod cells have the elongated flattened apical ends like rods, where numerous membranous disks are stacked in a parallel arrangement. These membranous disks contain rhodopsin, a purple visual pigment which is bleached by light and which starts visual activation. The number of the rod cells in the human retina can reach approximately 120 million. They are extremely sensitive to the light such that they are usually easily activated when dark environments are encountered. The cones also have elongated apical ends but shaped as cones, unlike the rod cells. At approximately six million, there are far fewer cone cells in the human retina compared to the number of rod cells. They also have numerous piled membranous disks with iodopsin, the cone photopigment, which reaches maximal sensitivity in the red, green, or blue region of the visible light spectrum. In addition, they are only sensitive to higher intensities of light than those enabling rod cells to be stimulated. Therefore, the main roles of cone cells in the visual function are color discrimination and the accomplishment of fine vision. Cone cells are distributed mainly over the central area of the retina, while rod cells are widely distributed over all areas of the retina. This fact tells us that at night, peripheral vision should be utilized for safety. At the posterior pole of the mid-visual axis of the retina, there is a pit structure, called the central fovea, where cone cells are more numerous, allowing the finest vision to be produced (Fig. 12.2).

The second layer is an intermediate layer of photosensitive cells consisting of bipolar neurons which connect rods and cones to ganglion cells. The final layer is an internal layer of these cells, composed of the ganglion cells with long dendrites and which form the optic nerve through their collective arrangement, through which signals of visual sensing are sent to the visual cortex.

Eye, Oxidative Damage (Including Sunlight), and Vitamin C

Sunlight plays the essential role in the visual function. It contains several electromagnetic rays, such as visible rays, infrared rays, and ultraviolet rays, among which only the visible rays (wavelength range: 380 ~ 770 nm) are indispensable for vision.

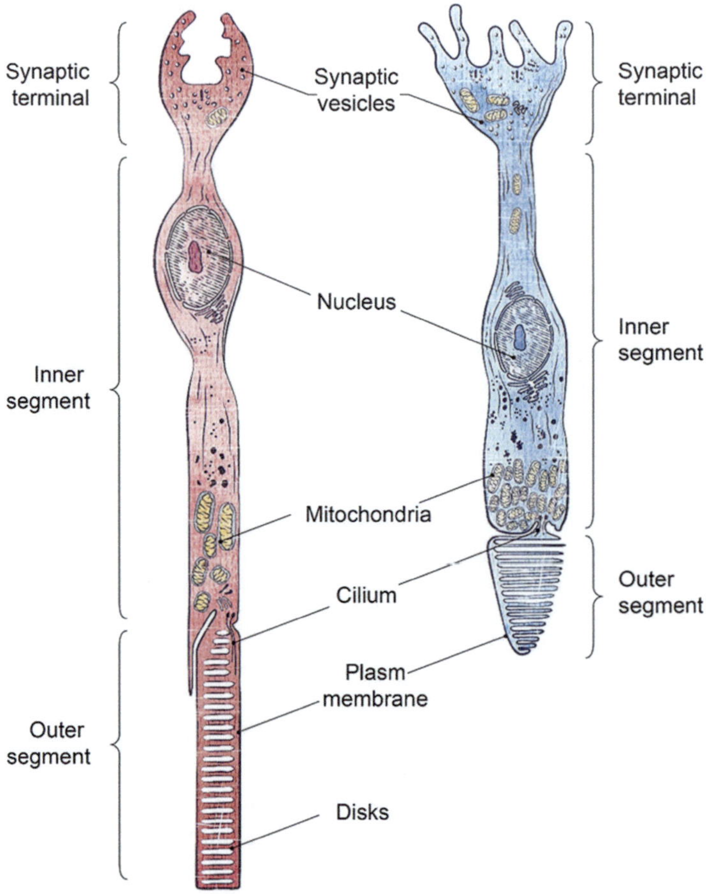

Fig. 12.2 Schematic illustration of photoreceptive cells, rod (left) and cone cells (right). Rod cells are responsible for recognition of light and darkness, whereas cone cells are responsible for recognition of color vision

However, ultraviolet (UV) rays are usually also transmitted through the ophthalmic refractive media during the visual function. UV radiation is one of many causes of oxidative damage. It is quite natural that chronic ophthalmic diseases such as cataracts and macular degeneration increase with age because the chronic damage caused by UV radiation accumulates with age.

Vitamin C level in the eye physiologically is found at rates of 10–40 times higher than in human plasma (50 μM). Vitamin C levels in aqueous humor and vitreous humor are 200–500 μM and 2 mM, respectively (Duarte and Lunec 2005). Vitamin C has diverse functions in the body as an antioxidant (Padayatty et al. 2003). Considering its anti-oxidative effects, the physiologically high concentration of vitamin C in the eye suggests that vitamin C plays an important role in the fight against age-related damage and oxidative stress in ocular diseases. The concentration of vitamin C in aqueous humor decreases with age in patients with age-related cataracts (Canadananovic et al. 2015). Vitreous concentrations of vitamin C are impaired in diabetic patients. In diabetic retinopathy, hypoxia-induced oxidative stress plays an important role in the development of ischemic retinopathy (Li et al. 2012). Metabolic mechanisms are associated with the excess production of reactive oxygen species (ROS) and the depletion of antioxidants in diabetic retinopathy (Yang et al. 2010). Serum levels of vitamin C are lower in patients with diabetes (Sinclair et al. 1991; Will and Byers 1996; Iino et al. 2005). As diabetic retinopathy progresses, the vitreous concentrations of vitamin C level are decreased. Takano et al. reported that vitamin C levels of vitreous in patients with proliferative diabetic retinopathy were lower than those in nondiabetic retinopathy patients (Takano et al. 1997). The high level of vitamin C in the vitreous is maintained by a sodium-dependent ascorbate transporter (SLC23A2) in the pigmented layer of the ciliary epithelium (Tsukaguchi et al. 1999). It can be presumed that SLC23A2 expression levels in diabetic patients may be suppressed, leading to the impaired accumulation of vitamin C in the vitreous.

Byshneva and Senchuk suggested that ascorbate may play an important role in UV-induced lens pathology in an in vitro study demonstrating the effect of UV radiation on the level of ascorbate, SH-groups, and glutathione reductase activity in the soluble fraction of bovine eye lenses (Byshneva and Senchuk 2002). Toskela et al. reported that the concentrations of vitamin C in the aqueous humor of diurnal species (*Acomys russatus*), which are usually exposed to much more sunlight than nocturnal species, were 35 times higher than those in the aqueous humor of nocturnal species (*Acomys cahirinus*) (Koskela et al. 1989). Varma et al. demonstrated that the in situ generation of active radicals and other active species of oxygen could induce lens damage and cataract formation in in vitro experiments with rat lenses cultured in a medium which generated oxygen radicals (Varma and Richards 1988). Williams and Paterson also found a protective role of ascorbic acid during inflammatory episodes in the eye in in vitro studies demonstrating that activated leukocytes reacted with ascorbic acid, diminishing the concentration of vitamin C in the surrounding medium (Williams and Paterson 1986).

Khordad et al. even reported that the administration of ascorbic acid with garlic acid during pregnancy and lactation may protect against lead-induced apoptosis in the eye retinas of rat offspring in an in vivo study using pregnant Wistar rats (Khordad et al. 2013).

Although there have been some reports concerning adverse effects of vitamin C on the eye from in vitro studies, unfortunately these groups did not realize that normally the aqueous humor contains substantial amounts of vitamin C, nor did they elucidate why such amounts of vitamin C were contained in this fluid and/or the biological significance of these amounts (Atalay et al. 1998; Avila et al. 2010).

References

Atalay A, Ogus A, Bateman O, Slingsby C. Vitamin C induced oxidation of eye lens gamma crystallins. Biochimie. 1998;80(4):283–8.

Avila F, Friguet B, Silva E. Simultaneous chemical and photochemical protein crosslinking induced by irradiation of eye lens proteins in the presence of ascorbate: the photosensitizing role of an UVA-visible-absorbing decomposition product of vitamin C. Photochem Photobiol Sci. 2010;9(10):1351–8.

Byshneva LN, Senchuk VV. Effect of UV-radiation on the level of ascorbic acid, SH-groups, and activity of glutathione reductase in the eye lens. Vopr Med Khim. 2002;48(5):455–60.

Canadananovic V, Latinovic S, Barisic S, Babic N, Jovanovic S. Age-related changes of vitamin C levels in aqueous humour. Vojnosanit Pregl. 2015;72(9):823–6.

Duarte TL, Lunec J. Review: when is an antioxidant not an antioxidant? A review of novel actions and reactions of vitamin C. Free Radic Res. 2005;39(7):671–86.

Iino K, Fukui T, Anzai K, Iwase M, Kogawa K, Ogimoto M, et al. Serum vitamin C levels in type 2 diabetic nephropathy. Diabetes Care. 2005;28(11):2808–9.

Khordad E, Fazel A, Ebrahimzadeh Bideskan A. The effect of ascorbic acid and garlic administration on lead-induced apoptosis in rat offspring's eye. Iran Biomed J. 2013;17(4):206–13.

Koskela TK, Reiss GR, Brubaker RF, Ellefson RD. Is the high concentration of ascorbic acid in the eye an adaptation to intense solar irradiation? Invest Ophthamol Vis Sci. 1989;30(10):2265–7.

Li SY, Fu ZJ, Lo AC. Hypoxia-induced oxidative stress in ischemic retinopathy. Oxidative Med Cell Longev. 2012;2012:426769.

Padayatty SJ, Katz A, Wang Y, Eck P, Kwon O, Lee JH, et al. Vitamin C as an antioxidant: evaluation of its role in disease prevention. J Am Coll Nutr. 2003;22(1):18–35.

Sinclair AJ, Girling AJ, Gray L, Le Guen C, Lunec J, Barnett AH. Disturbed handling of ascorbic acid in diabetic patients with and without microangiopathy during high dose ascorbate supplementation. Diabetologia. 1991;34(3):171–5.

Takano S, Ishiwata S, Nakazawa M, Mizugaki M, Tamai M. Determination of ascorbic acid in human vitreous humor by high-performance liquid chromatography with UV detection. Curr Eye Res. 1997;16(6):589–94.

Tsukaguchi H, Tokui T, Mackenzie B, Berger UV, Chen XZ, Wang Y, et al. A family of mammalian Na^{+-} dependent L-ascorbic acid transporters. Nature. 1999;399(6731):70–5.

Varma SD, Richards RD. Ascorbic acid and the eye lens. Ophthalmic Res. 1988;20(3):164–73.

Will JC, Byers T. Does diabetes mellitus increase the requirement for vitamin C? Nutr Rev. 1996;54(7):193–202.

Williams RN, Paterson CA. A protective role for ascorbic acid during inflammatory episodes in the eyes. Exp Eye Res. 1986;42(3):211–8.

Yang Y, Hayden MR, Sowers S, Bagree SV, Sowers JR. Retinal redox stress and remodeling in cardiometabolic syndrome and diabetes. Oxidative Med Cell Longev. 2010;3(6):392–403.

Chapter 13
Summary and Conclusion

Every animal produces vitamin C by themselves except for guinea pigs and primates including humans.

The daily amount of vitamin C production in each animal is in grams, not milligrams, when converted to human body weight.

In the urine of animals that synthesize vitamin C by themselves, there is 3–5 times more vitamin C compared to its blood concentration.

In addition, there is a sophisticated device (SVCT1 expression) that controls the urinary excretion of vitamin C in the proximal tubules of the kidneys, suggesting that the presence of vitamin C in the urine of the animal is not excretion but must have a certain biological role, which seems to be intended to protect the bladder mucosa by removing the oxygen free radicals in the urine.

These facts indicate that the criteria for determining the daily optimal allowance to date should be reconsidered.

Vitamin C is a representative antioxidant, and its radical, the ascorbyl radical, which is formed after its antioxidant function, does have the least toxicity, while the radicals generated from other antioxidants after their antioxidant function are usually very toxic. Furthermore, vitamin C can regenerate toxic radicals generated from other antioxidants after their antioxidant function to the original antioxidants, suggesting that people should not take other antioxidants only but take them with a high dose of vitamin C.

In addition to the antioxidant function, vitamin C has a very important role as a cofactor in eight important enzymatic reactions that take place in the human body: wound healing process ((1) prolyl and (2) lysyl hydroxylases), a critical role in energy metabolism ((3) and (4) two dioxygenases), a critical function in the biosynthesis of norepinephrine ((5) dopamine-β-monooxygenase), a role in cholesterol metabolism ((6) 7α-monooxygenase), a role in downregulating high blood pressure ((7) endothelial nitric oxide synthase), and an important role in the ability of cells to recognize and respond to hypoxia ((8) oxygen-dependent prolyl hydroxylase).

Vitamin C has very important roles in protecting all organs constituting the human body from oxidative damages as well as promoting reactions involved in

W. J. Lee, *Vitamin C in Human Health and Disease*,
https://doi.org/10.1007/978-94-024-1713-5_13

several life processes, resulting in the following functions: prevention of atheroma formation by inhibiting injury of the arterial endothelia and lipid peroxidation; brain tissue protection from oxidative damage (highest concentration of vitamin C in the neurons: 5–10 mM, 100–200 times as high as the peripheral blood concentration); antiviral function directly or indirectly by immune enhancement; antiaging function by delaying frailty; hepatic protection function by immune modulation; anticancer functions, killing cancer cells directly (a high dose of vitamin C higher than 5 mM achieved only by IV administration) or preventing cancer formation through destruction of immune escape operated by cancer cells (a low dose of vitamin C; 200–500 μM achieved by oral administration); eye protection against sunlight (UV irradiation), resulting in a delay of cataract formation or retinal degeneration; anti-aging functions caused by sunlight in the skin, including whitening of the skin, delay of the wrinkle formation, and so on; functions reducing the somatic and mental damages caused by prolonged stress, which is a fate of modern people; and even preventing people from experiencing a sudden death.

The pharmacokinetic study of vitamin C in the humans tells us that it has a 6 h cycle in the human peripheral blood, indicating specifically that the vitamin C level in the blood reaches its peak 3 h after oral intake with a meal and, then, it comes back to the original basic level 6 h after oral intake.

Therefore, it is most desirable to take vitamin C every 3 h, but it is practically impossible to do in practice. Instead, it is recommended to take it every 6 h, actually with every meal (people usually have three meals at almost 6 h intervals).

Two more reasons why people should take vitamin C with their meals (more strictly during meals) are because people may suffer from stomach pain if they take it on an empty stomach, and vitamin C inhibits the production of a well-known carcinogen, nitrosamine, that occurs during mixing in the stomach.

Vitamin C is a drug that is safer than any medicine in the world. Even with high doses, there are almost no side effects. However, one adverse effect to be aware of is that people who have ever previously suffered from urinary stones may have them more frequently. Those people are strongly recommended to drink plenty of water during their daily lives.

In conclusion, people are strongly recommended to take 2000 mg of vitamin C every meal. Eventually, they will take a total of 6000 mg of vitamin C per day.

MIX
Papier aus verantwortungsvollen Quellen
Paper from responsible sources
FSC® C105338

If you have any concerns about our products,
you can contact us on
ProductSafety@springernature.com

In case Publisher is established outside the EU,
the EU authorized representative is:
Springer Nature Customer Service Center GmbH
Europaplatz 3, 69115 Heidelberg, Germany

Printed by Libri Plureos GmbH
in Hamburg, Germany